Lecture Notes in Earth Sciences   68

Editors:
S. Bhattacharji, Brooklyn
G. M. Friedman, Brooklyn and Troy
H. J. Neugebauer, Bonn
A. Seilacher, Tuebingen and Yale

Springer-Verlag Berlin Heidelberg GmbH

Peter Stille   Graham Shields

Radiogenic
Isotope Geochemistry
of Sedimentary and
Aquatic Systems

With 144 Figures and 14 Tables

 Springer

Authors

Dr. Peter Stille
Research Director
Centre de Geochimie de la Surface (C.N.R.S.)
1 rue Blessig, Strasbourg Cedex, France

Dr. Graham Shields
Centre de Geochimie de la Surface (C.N.R.S.)
1 rue Blessig, Strasbourg Cedex, France

"For all Lecture Notes in Earth Sciences published till now please see final pages of the book"

Library of Congress Cataloging-in-Publication Data

Stille, Peter, 1950-
    Radiogenic isotope geochemistry of sedimentary and aquatic systems
/ Peter Stille, Graham Shields.
       p.. cm. -- (Lecture notes in earth sciences, ISSN 0930-0317 ;
68)
    Includes bibliographical references and index.

    1. Isotope geology. 2. Weathering. 3. Sediments (Geology)
I. Shields, Graham, 1970-   . II. Title. III. Series.
QE501.4.N9S75 1997
551.9--dc21   97-34436
                                                        CIP

ISSN 0930-0317
ISBN 978-3-540-63177-4   ISBN 978-3-540-69208-9 (eBook)
DOI 10.1007/978-3-540-69208-9

© Springer-Verlag Berlin Heidelberg 1997
Originally published by Springer-Verlag Berlin Heidelberg New York in 1997.

Typesetting: Camera ready by author
SPIN: 10528848   32/3142-543210 - Printed on acid-free paper

Preface

This book is based on the script of a lecture course in isotope geochemistry, which is given at the University of Strasbourg, France and at the Eidgenössische Technische Hochschule (ETH), Zurich, Switzerland by Peter Stille. It is intended to be read by geologists, hydrologists, geochemists and any researchers and students from the broad field of environmental science. Its purpose is to enable readers to venture safely into the often mirky realms of radiogenic isotope geochemistry applied to sedimentary and aquatic systems. The authors have strived to construct the book in such a way that it can be read and understood by those readers, especially students, who have no background in geochemistry. 7 chapters take the reader through the rock cycle from weathering on the continent to eventual deposition in the sea, looked at largely from the perspective of radiogenic isotope geochemistry.

Isotope geochemistry has become the most important field of speciality in geochemistry and is being applied increasingly to the study of chemical processes in natural water bodies, the atmosphere and to other aspects of our environment. By integrating various isotope systems we can demonstrate that chemical exchange or movement of material has taken place in, for example, waste dumps or during natural sedimentation processes, (including weathering, erosion and diagenesis) as a result of water-rock exchange. The study of isotope systems can yield important information which allows us to determine the origin of migrating fluid phases, the water-rock ratio necessary for such exchange and the mechanisms which have led to the mobility of elements.

This book attempts to shed some light on the entire field of sedimentary isotope geochemistry, including the isotope geochemistry of natural water bodies, and objectively discusses important results of previous research in the various sub-fields. Readers are thus given the opportunity to critically assess this research, allowing them to apply isotopes in their own particular fields. In order to do this, we will deal especially with the geochemical and environmental questions which can be and have been solved through the application of an isotopic study.

The sequence of chapters reflects the rock cycle. Weathering and erosion contribute significantly to sediment formation and are discussed in chapter 2.

In chapter 3, the fluvial transport of these weathering products into the sea is looked at in some detail. The rivers transport large quantities of chemical elements into the ocean basins either in dissolved form or associated with the suspended load. Knowledge of these transport mechanisms allows us to establish mass balances for the oceans. Isotope studies help us also to reconstruct chemical exchange processes between the suspended load, river water and river basin sediments as well as determining the origin of toxic elements. Isotope geochemistry comes into its own when used to identify the sources of contamination in river water and to trace the transport routes of unwelcome chemical substances.

Chapter 4 concerns the application of various isotopic systems to the investigation of our heavily polluted biosphere. It is designed especially to encourage environmental researchers to explore the application of isotope geochemistry to their studies. Problems of source, mobility and exchange characteristics of heavy metals in contaminated soils and natural water bodies are discussed here.

Chapters 5, 6 and 7 are all concerned with the marine environment. The oceans represent the most important sedimentary environment. River loads end up here and authigenic minerals (e.g. phosphates, carbonates, clay minerals, FeMn-ore bodies) are formed here. Knowledge of the isotopic composition of oceans today and in the past is of the greatest importance not only for understanding the processes of exchange between sediments and seawater and during diagenesis but also for the reconstruction of the paleo-environment. Residence times of elements in seawater are explained in Sect. 5.3. Elements with long residence times (relative to the ocean mixing time) can be used for high resolution chemostratigraphy (e.g. Sr) whereas those with short residence times can be used to trace ocean currents and reconstruct ocean circulation patterns in the past (e.g. Nd, Pb, Ce, O). As well as looking in detail at the application of the Rb-Sr and the Sm-Nd systems to stratigraphy and paleoceanography, this chapter introduces less intensely researched isotope systems such as the U-Th-Pb, Re-Os and La-Ce systems and their potential for future marine research.

While chapter 5 concentrates on seawater and authigenic mineral isotope geochemistry, chapter 6 deals more especially with detrital and authigenic clay minerals in marine sediments for dating and for constraining mechanisms of exchange during deposition and diagenesis. The Rb-Sr and K-Ar isotope systems of clay-rich rocks are explored here. Special attention is paid to the formation and dating potential of glauconite.

Chapter 7 concerns the behaviour of the Sm-Nd isotope system in marine sediments during diagenesis. Chemical and isotopic exchange mechanisms leading to Nd isotopic homogenization between authigenic clay minerals in phosphate-rich, detrital sediments and bituminous shales are discussed. Knowledge of these mechanisms is of great importance, not only for the reconstruction of diagenetic evolution in a sedimentary basin, but also for dating these diagenetic processes. However, as was shown recently, knowledge of the Sm-Nd isotopic system can also be important for petroleum exploration, since neodymium isotopic equilibrium may occur between authigenic clay minerals, ambient waters, source rocks and petroleum.

We have benefited from many discussions with, and helpful comments from, several of our colleagues and friends. Peter Stille would like to express his gratitude to Emilie Jäger (Bern) and Mitsunobu Tatsumoto (Denver) who were undoubtedly the greatest influences during his formative years in science. Thanks go to Norbert Clauer (Strasbourg) for first introducing Peter Stille to sediments. We are also grateful to François Gauthier-Lafaye, Regis Bros, Marc Steinmann, Miriam Andres, Urs Schaltegger, Horst Zwingmann and Jost Eikenberg. Especial thanks go to Claude Hammel who drafted all the figures.

Strasbourg, France   Peter Stille
July 1997   Graham Shields

Table of Contents

1 Introduction and Basic Principles of Isotope Geochemistry

This book does not pretend to be complete. For example, we do not consider stable isotopes in any great depth, but instead concentrate on the radiogenic isotopic systems such as Rb-Sr, Sm-Nd, U-Pb, etc. Stable isotope systems are dealt with in more detail in the books of Kyser (1987) and Arthur and Anderson eds. (1983). Special topics within the field of sedimentary isotope geochemistry are dissected and discussed in the book "Isotopic signatures and sedimentary records" (Clauer and Chaudhuri 1992). Information about the dynamics of isotope systems in clays is given in the book "Clays in crustal environments" (Clauer and Chaudhuri 1995). Various applications of the uranium decay series for the study of sedimentary and aquatic systems are not covered in this book as they are already dealt with in a comprehensive text book (Ivanovich and Harmon 1992). This first chapter is concerned with some of the most important basic parameters in isotope geochemistry and will help in the reading and understanding of the chapters that follow. A more comprehensive treatment with detailed references is given in G. Faure's standard work, "Principles of Isotope Geology".

1.1 Stable Isotopes

It is generally understood that the isotopic composition of the stable isotopes is represented as the relative enrichment of the heavier to the lighter isotope. In order to represent the isotopic composition of hydrogen, the deuterium (D= ^{2}H) - hydrogen ratio (D/H) is used. Similarly, the ratio $^{18}O/^{16}O$ is used to represent the isotopic composition of oxygen; the ratio $^{13}C/^{12}C$ for carbon and the ratio $^{34}S/^{32}S$ for sulfur. All these isotopic compositions are commonly presented as isotopic enrichments relative to internationally recognized standards:

$$\delta^{18}O = \left[\frac{(^{18}O/^{16}O)_{sample} - (^{18}O/^{16}O)_{SMOW}}{(^{18}O/^{16}O)_{SMOW}} \right] * 10^3$$

(1)

(where SMOW: "Standard Mean Ocean Water" for oxygen und hydrogen. Other standards are PDB: Belemnite from the Cretaceous Pee Dee Formation (USA) for carbon and oxygen. CDT: Canyon Diablo Troilite, which is an iron sulphide meteoritic mineral for sulfur).

Due to their different physico-chemical characteristics, light or heavy isotopes may be relatively concentrated in some molecules and mineral phases. Evaporation, condensation and freezing of water can lead to fractionation of O and H isotopes, whereas biological processes profoundly influence the fractionation of C and to a certain extent S isotopes in the natural environment. This process of fractionation can be represented by an isotopic fractionation factor, α. For example:

$$\alpha = \frac{R_A}{R_B}$$

(II)

where $R_A = (^{18}O/^{16}O)$ in phase A and $R_B = (^{18}O/^{16}O)$ in phase B. The following relationship exists between the fractionation factor and temperature (T):

$$1000 \ln \alpha = A \left(10^6 T^{-2} \right) + B = \delta A - \delta B$$

(III)

where A, B are constants. If, for example, two solid phases have undergone oxygen exchange with a common reservoir at a particular temperature, the difference in their $\delta^{18}O$ values is a function of this temperature. This rule allows us to reconstruct the equilibrium temperatures at which rock forming minerals have formed on the basis of oxygen isotopic compositions.

1.2 Radiogenic Isotopes

The radiogenic isotopic composition of a element is shown as the ratio of the abundance of a radiogenic isotope to a non-radiogenic isotope of that element. The following radiogenic isotopic ratios are used internationally: $^{40}Ar/^{36}Ar$, $^{87}Sr/^{86}Sr$, $^{206}Pb/^{204}Pb$, $^{207}Pb/^{204}Pb$, $^{208}Pb/^{204}Pb$, $^{143}Nd/^{144}Nd$ for the elements Argon, Strontium, Lead and Neodymium, where ^{40}Ar, ^{87}Sr and ^{143}Nd represent the radiogenic isotopes. The abundances of radiogenic isotopes in rocks are very variable and dependent on the amount originally present of the "parent" isotope and on the decay constant. The general equation for radioactive decay can be written as:

$$N = N_o e^{-\lambda t} \tag{IV}$$

where "N" is the number of radioactive parent atoms at any time "t", "N_0" is the original number of atoms at time "t=0" and "λ" is the decay constant. However, equation IV does not permit us to date geological substances directly as we do not know "N_0", i.e. the number of parent atoms at the start of the chronometer. Equation IV may be manipulated for this purpose in the following way:

$$D = N_o - N = N e^{\lambda t} - N = N\left(e^{\lambda t} - 1\right) \tag{V}$$

where D corresponds to the number of radiogenic daughter atoms present.

1.2.1 The Rb-Sr Method

The Rb-Sr isotope system can be used to determine the age of rocks and minerals because of the constancy of the law of radioactivity. Strontium has four naturally occurring isotopes: ^{84}Sr, ^{86}Sr, ^{87}Sr und ^{88}Sr. Rubidium has two naturally occurring isotopes: ^{87}Rb and ^{85}Rb. The strontium isotopic signature of rocks and minerals is represented by the ratio $^{87}Sr/^{86}Sr$. ^{87}Rb is the parent isotope and undergoes radioactive decay to produce ^{87}Sr by emission of a negative β-particle. The radioactive decay constant of ^{87}Rb, λ_{Rb}, is $1.42 * 10^{-11}y^{-1}$ (y:year).

Let us use our standard equations again. If we apply equation (V) and include in this a normalization in order to reconstruct both the original composition and the increase of $^{87}Sr/^{86}Sr$ through time using the non-radiogenic, stable ^{86}Sr isotope for this purpose, we can produce the following equation:

$$^{87}Sr \, /^{86} \, Sr =^{87} Rb \, /^{86} \, Sr\left(e^{\lambda t} - 1\right) \tag{VI}$$

However, this is not yet sufficient for us to determine the age of a rock. A rock or mineral system originally contains a certain amount of strontium already and therefore radiogenic ^{87}Sr, too. At its origin, this system must have possessed a lower, inital $(^{87}Sr/^{86}Sr)_i$ isotopic ratio (see also Sect. 1.3). Equation (VI) may thus be expanded:

$$(^{87}Sr \, /^{86} \, Sr)_{now} = (^{87}Sr \, /^{86} \, Sr)_i + (^{87}Rb \, /^{86} \, Sr)\left(e^{\lambda t} - 1\right) \tag{VII}$$

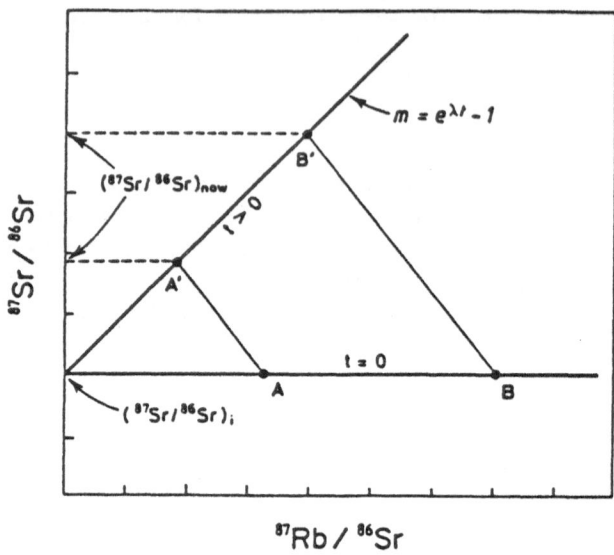

$^{87}Rb / {}^{86}Sr$

Fig. 1.1. Schematic Rb-Sr isochron diagram. Cogenetic mineral phases A and B formed at the same time (t=0) and integrated the same initial ($^{87}Sr/^{86}Sr$)i ratio but different quantities of Rb and Sr. Since the time of crystallization, the radioactive decay of ^{87}Rb has produced radiogenic ^{87}Sr leading to A' and B' (now). Both points define a straight line (isochron) whose intercept with the $^{87}Sr/^{86}Sr$ axis corresponds to the initial Sr ratio. The slope of the line ($e^{\lambda t}$-1) is equal to the time elapsed since the formation of mineral phases A and B.

This equation corresponds to a straight line of the general form :Y= B+X*m. Axes Y and X represent $^{87}Sr/^{86}Sr$ and $^{87}Rb/^{86}Sr$, respectively. The factor ($e^{\lambda t}$-1) corresponds to the steepness or gradient of the straight line in Fig. 1.1. The gradient is thus dependent on the time passed since the beginning of the chronometer. At time "t=0", the gradient is zero. The older the system, the steeper the straight line. Therefore, this gradient allows us to determine the age directly. Equation (VII) can be reduced to the general equation for radioactive age dating:

$$ t = \frac{1}{\lambda} \ln\left(\frac{\left({}^{87}Sr / {}^{86}Sr \right)_{now} - \left({}^{87}Sr / {}^{86}Sr \right)_{i}}{\left({}^{87}Rb / {}^{86}Sr \right)} + 1 \right) \qquad \text{(VIII)} $$

If the inital strontium isotopic composition of the system is unknown, we can only determine the age if there are several components of this system present, which formed in equilibrium with each other and had different inital Rb/Sr ratios. If these mineral components formed at the same time, then we can define a so-called isochron, which would have been a straight, horizontal line at the time of formation (isotopic equilibrium, isotopic homogenization). After time "t>0" a straight line will be formed which is also termed an isochron but whose gradient corresponds to the increase in the value of $(e^{\lambda t}-1)$. From the gradient of this straight line we can work out the time at which the system was in isotopic equilibrium assuming that the system has remained closed in the meantime. Subsequent hydrothermal events, metamorphosis or diagenesis can lead to the reopening of the system and to a rehomogenization of the isotopic ratios. When the newly opened system recloses after such an event, our clock starts again, this time with a higher, initial isotopic ratio.

1.2.2 The U-Th-Pb Method

The parent isotopes ^{238}U (Uranium), ^{235}U and ^{232}Th (Thorium) undergo decay by way of a long sequence of half-way houses to form the stable lead isotopes ^{206}Pb, ^{207}Pb and ^{208}Pb, respectively. The ^{204}Pb isotope is not produced through radioactive decay. The time-dependent evolution of lead isotopes can be represented by the following equations:

$$\left(^{206}Pb/^{204}Pb\right)_{now} = \left(^{206}Pb/^{204}Pb\right)_i + \left(^{238}U/^{204}Pb\right)\left(e^{\lambda_1 t}-1\right) \qquad \text{(X)}$$

$$\left(^{207}Pb/^{204}Pb\right)_{now} = \left(^{207}Pb/^{204}Pb\right)_i + \left(^{235}U/^{204}Pb\right)\left(e^{\lambda_2 t}-1\right) \qquad \text{(XI)}$$

$$\left(^{208}Pb/^{204}Pb\right)_{now} = \left(^{208}Pb/^{204}Pb\right)_i + \left(^{232}Th/^{204}Pb\right)\left(e^{\lambda_3 t}-1\right) \qquad \text{(XII)}$$

where λ_1, λ_2 und λ_3 are the radioactive decay constants for ^{238}U, ^{235}U und ^{232}Th.

$\lambda_1(^{238}U)= 1.55125*10^{-10}/y$; $\lambda_2(^{235}U)= 9.8485*10^{-10}/y$; $\lambda_3(^{232}Th)=4.9475*10^{-11}/y$

reducing equations X und XI gives:

$$\frac{\left(^{207}Pb/^{204}Pb\right)_{now} - \left(^{207}Pb/^{204}Pb\right)_i}{\left(^{206}Pb/^{204}Pb\right)_{now} - \left(^{206}Pb/^{204}Pb\right)_i} = \frac{^{235}U\left(e^{\lambda_2 t}-1\right)}{^{238}U\left(e^{\lambda_1 t}-1\right)} \qquad \text{(XIII)}$$

where the $^{235}U/^{238}U$ ratio for terrestrial material with a normal isotopic composition is constant (1/137.88).

1.2.3 The Sm-Nd Method

Samarium and neodymium belong to the group of elements called the rare earth elements (REE) and both possess 7 isotopes (^{144}Sm, ^{147}Sm, ^{148}Sm, ^{149}Sm, ^{150}Sm, ^{152}Sm, ^{154}Sm; ^{142}Nd, ^{143}Nd, ^{144}Nd, ^{145}Nd, ^{146}Nd, ^{148}Nd, ^{150}Nd). ^{147}Sm decays through the emission of an alpha particle to the isotope ^{143}Nd. The decay constant equals $6.54 * 10^{-12}y^{-1}$. The Nd isotopic composition of rocks and minerals is presented using ^{143}Nd/^{144}Nd. The increase in radiogenic Nd content through time can be represented as follows:

$$\left(^{143}Nd\,/^{144}\,Nd\right)_{now} = \left(^{143}Nd\,/^{144}\,Nd\right)_i + \left(^{147}Sm\,/^{144}\,Nd\right)\left(e^{\lambda t}-1\right) \quad (XIV)$$

To produce the age equation, see equation (VIII). Frequently, the isotopic composition is shown normalized to the commonly used chondritic meteorite standard in the form of εNd values:

$$\varepsilon Nd_{now} = \left(\frac{\left(^{143}Nd\,/^{144}\,Nd\right)_{sample}}{\left(^{143}Nd\,/^{144}\,Nd\right)_{CHUR}}-1\right)*10^4 \quad (XV)$$

where CHUR: chondritic uniform reservoir. Today's ratios are:
^{143}Nd/^{144}Nd : 0.51264; ^{147}Sm/^{144}Nd : 0.1967

1.2.4 The K-Ar Method

3 Argon isotopes, ^{36}Ar, ^{38}Ar und ^{40}Ar occur in nature. The isotope ^{40}Ar is radiogenic and is formed through the radioactive decay of ^{40}K. ^{40}K also undergoes radioactive decay to ^{40}Ca. The increase through time in the amounts of radiogenic ^{40}Ar und ^{40}Ca may be represented using the following equation:

$$^{40}Ar + ^{40}Ca = ^{40}K\left(e^{\lambda t}-1\right) \quad (XVI)$$

where λ represents the decay constant for ^{40}K. λ is produced by using the decay constants for the decay of ^{40}K to ^{40}Ar (λ_e) and ^{40}K to ^{40}Ca (λ_b) together:

$\lambda = \lambda_e + \lambda_b$, where
$\lambda_e = 0.581 * 10^{-10}y^{-1}$
$\lambda_b = 4.962 * 10^{-10}y^{-1}$
$\lambda = 5.543 * 10^{-10}y^{-1}$

The time dependent increase in the amount of radiogenic ^{40}Ar can be represented using the following equation:

$$\left(^{40}Ar\right)_{rad} = \frac{\lambda e}{\lambda}\left(^{40}K\right)\left(e^{\lambda t}-1\right) \quad (XVII)$$

1.3 Pb, Sr and Nd Isotopes as Geochemical "Tracers"

Unlike the stable isotopes, H, C or O, which can undergo isotopic fractionation depending on temperature or nature of the chemical reaction (see Sect.1.1), Pb, Sr and Nd do not appear to show any such behaviour due to their significantly higher atomic masses. For example, the $^{87}Sr/^{86}Sr$ ratio of a mineral is solely dependent upon its Rb/Sr ratio, the time of formation and the initial $^{87}Sr/^{86}Sr$ ratio (see Sect.1.2.1).

The inital isotopic ratio which characterizes the beginning of the chronometer ought to be identical to that of the fluid phase in which the mineral crystallized. Therefore, this ratio can help us determine the origin, environment of formation and in certain circumstances the conditions of formation of a mineral. Similarly, knowledge of the initial isotopic ratio of a granite body can help us identify the source of the magma. As the mantle and the continental crust have different isotopic ratios, it is possible to differentiate between granites of mantle and crustal origin. Such source determinations are only possible because the isotopic ratios of Pb, Sr and Nd vary sufficiently within the Earth and because these ratios vary within characteristic end members that define specific, geochemically different reservoirs. The reasons for these characteristic reservoir isotopic signatures lie in various geochemical processes of the past which have led to elemental fractionation, for example, where U-Pb, Rb-Sr und Sm-Nd parent-daughter fractionation has taken place. The most important elemental fractionation took place during the development of the mantle-crust. This can be represented using a simplified Sm-Nd evolution model for the Earth (Fig. 1.2).

It is assumed today that the Earth, together with the other planets of the solar system, were born out of the same solar nebula and that the stony meteorites that periodically rain down on the Earth record the isotopic characteristics of this original nebula. Age dating which has been carried out on these meteorites (chondrites, achondrites, shergottites) suggests that the solar system orignated around 4580 million years ago. Because the Earth developed out of the same nebula, it is assumed that the Earth contained the same initial isotopic ratios as the meteorites at the time of their formation (A in Fig. 1.2). At this stage, the "chondritic" Earth ball was still primitive, geochemically speaking, unfractionated and isotopically homogeneous. Geochemical processes, which led to the formation of first stable, continental crust may have taken place about 3800 million years ago at the latest.

Enrichment and depletion of certain elements in the newly formed mantle and crust segments led to fractionation in the Rb-Sr und Sm-Nd parent-daughter systems (B in Fig. 2.2). Basically, that element of the parent-daughter system which had the greater ionic radius came to be preferentially enriched in the continental crust. This means that the parent rubidium became enriched over the daughter strontium and the daughter neodymium over the parent samarium. Therefore, the $^{143}Nd/^{144}Nd$ ratio in the continental crust was not allowed to rise as

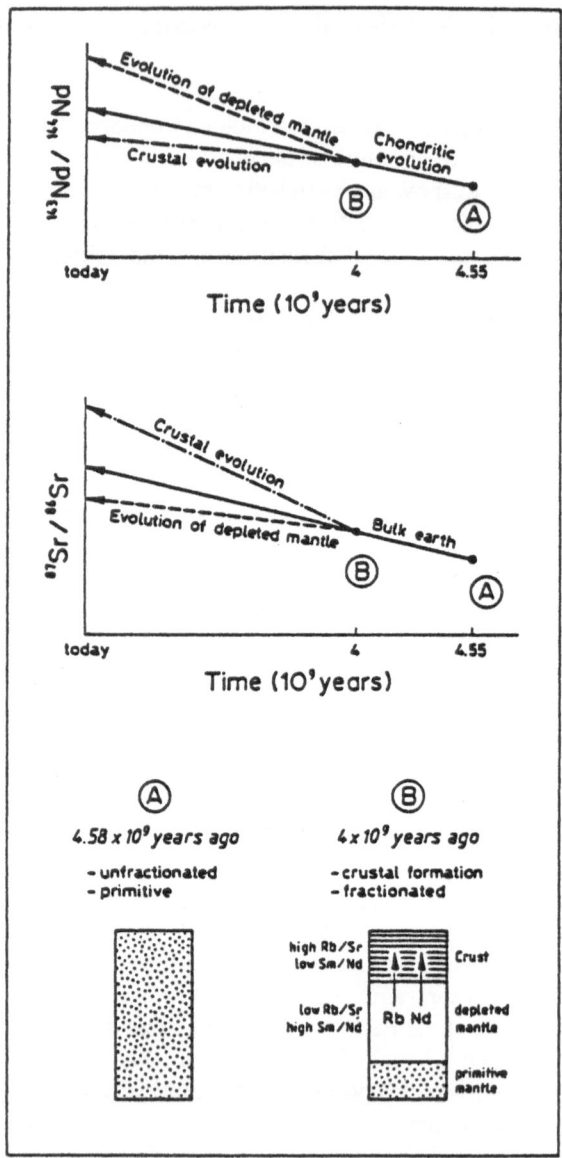

Fig. 1.2. Illustration of various Sr and Nd isotopic evolution parameters for the Earth discussed in the text.

much as its $^{143}Nd/^{144}Nd$ counterpart in the mantle, whereas in the Rb-Sr system the opposite was the case. As more rubidium is contained within the crust, more radiogenic ^{87}Sr is produced in the crust than in the mantle. Therefore, the crust shows a higher $^{87}Sr/^{86}Sr$ ratio than the mantle.

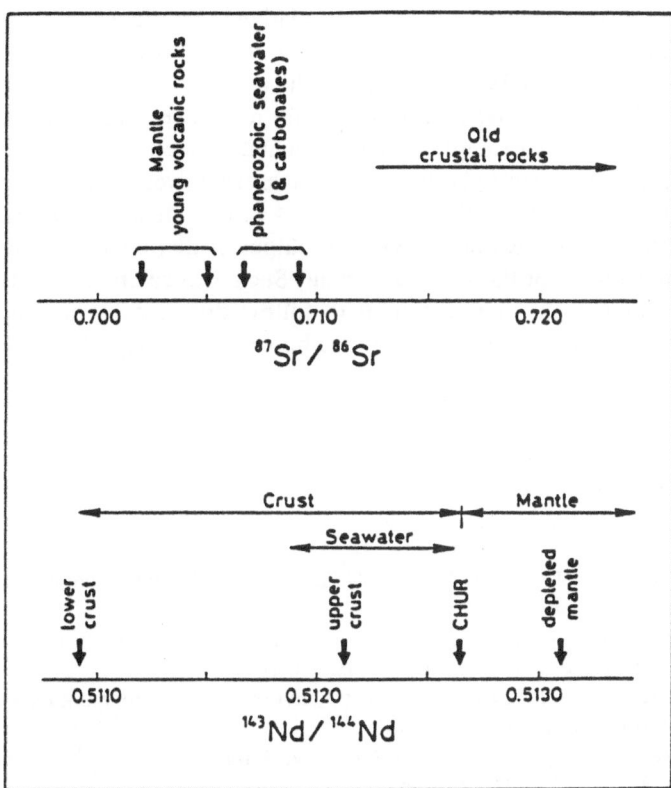

Fig. 1.3. Sr and Nd isotopic compositions of major geochemical reservoirs of the Earth.

As this once chondritic mantle is now depleted in typical crustal elements, one speaks of the mantle as a "depleted mantle" rather than as chondritic or primitive.

Uninterrupted, long running, geochemical, endogenic and exogenic processes of fractionation have allowed the development of a variety of different geochemical reservoirs during the long history of the Earth, with likewise characteristic isotopic signatures. In Fig. 1.3, the Sr and Nd isotopic ratios of the more important geochemical reservoirs are shown.

1.4 Elemental Residence Times

Knowledge of the residence time of an element is of great importance to understanding how rapidly the concentration of an element and its isotopic ratio will react to perturbations in the system. The residence time is defined as the concentration of that element in a medium relative to the fluxes in and out of the medium. Residence times can be calculated in marine or lake environments, in groundwater (see Sect. 4.2), in estuaries, in the atmosphere or for elements in the Earth's crust during tectonic recycling (crustal residence times) but the concept remains the same. For example, if an elemental residence time in the marine environment is calculated to be longer than the ocean water mixing time (about 3000 years), this means that the isotopic ratios of that element in seawater react only sluggishly to changes in the input or output of that element. This leads to isotopic homogeneity of the whole reservoir. Such a calculation is carried out in Sect. 5.2.1. for Nd and it is essential to follow this step by step in order to appreciate what residence time signifies and how it can be helpful.

1.5 References

Arthur MA, Anderson TF (eds)(1983) Stable isotopes in sedimentary geology. Soc. econ. petrog. min. Short course no. 10

Clauer N, Chaudhuri S (eds)(1992) Isotopic signatures and sedimentary records. Lecture Notes in Earth Sciences, Springer, Berlin Heidelberg New York

Clauer N, Chaudhuri S (1995) Clays in crustal environments. Isotope dating and tracing. Springer, Berlin Heidelberg New York

Faure G (1986) Principles of Isotope Geology. John Wiley and Sons, New York, Chichester, Brisbane, Toronto, Singapore

Ivanovich M, Harmon RS (eds)(1992) Uranium-series Disequilibrium. Oxford Science Publications, Calendron Press, Oxford

Kyser TK (ed)(1987) Short Course in Stable Isotope Geochemistry of Low Temperature fluids. Mineralogical Association of Canada, vol 13, Saskatoon.

2 Weathering

Even before the actual formation of sediments, processes were in operation that are vital to our understanding of sediment genesis. Particularly important in this respect are erosion and chemical weathering which reflect both the exchange with and the reaction of a rock to its environment, i.e. the atmosphere, biosphere or hydrosphere. Mechanical erosion and chemical weathering represent the beginnings of the sedimentary cycle on the surface of the Earth and result from changes in the thermodynamic conditions within a rock or mineral. For example, this can happen when crystalline rocks, which formed under extreme conditions of high temperature and pressure have to adapt to the wholly new physico-chemical conditons that exist close to the Earth's surface.

Weathering is a low temperature process. Chemical weathering of rocks is a consequence of the breaking down of the crystal lattice of a mineral. The type of weathering products and the rate of weathering are dependent on a number of factors such as mineralogic composition, topography, climate and vegetation. Since the beginnning of industrialization, acid rain and steadily increasing concentrations of pollutants in the atmosphere, soil and natural waters have also influenced the course of weathering.

The chemical reactions that take place during mineral weathering can be markedly different. It can generally be said that chemical weathering encompasses constructive as well as destructive mechanisms. Some minerals may go completely into solution, whereas others may leave behind one or a suite of new authigenic minerals such as iron oxide, carbonate or one of several clay minerals. Silicate minerals react with various sensitivities to the weathering process. Biotites and feldspars react much quicker than quartz and muscovite, an observation which had already been made by Goldich in 1938. Quartz and muscovite are difficult to dissolve and are frequently only abraded and disintegrated. It is important to recognize that the chemical composition of water in the weathering milieu is very strongly dependent on the types of minerals being weathered.

In what follows, we shall discuss the behaviour of various isotope systems of rocks and minerals during the weathering process. An important question will be, to what extent can isotopic equilibrium be reached in weathering profiles, allowing weathering and soils to be dated accurately.

2.1 Weathering of the Whole Rock

Dasch (1969) carried out the first comprehensive study of the behaviour of the Rb-Sr isotopic system of the whole rock during weathering. Among other things he investigated the course of weathering at various granite plutons including the Elburton granite of Piedmont and Georgia (USA), which can be found weathered to great depths, and tried not only to throw light on the mechanisms that take place during weathering but also the influence of weathering on the isotopic age. He described the various stages of weathering on the way to kaolinite. Sample material was collected from a profile where both fresh and weathered material was available. The geochemical profile demonstrated that Sr was preferentially dissolved from the original rock. The fresh granite contained 290 ppm Sr, whereas the most strongly weathered material contained only 84 ppm Sr. Rb content also decreased with increasing weathering but to a lesser extent than with Sr. It follows therefore that Rb/Sr ratios increase markedly with increasing weathering.

Thus, only few weathered samples lie on or near the isochron for unweathered samples that would date the age of intrusion of the granite pluton (Fig. 2.1). Strongly weathered rocks lie to the right of the age-reference line. As a consequence, weathering results in a younging of the Rb-Sr age, as the gradient of the reference line decreases. Another consequence of weathering, apart from the rise in Rb/Sr ratio, can be that the $^{87}Sr/^{86}Sr$ ratio will rise, too. The strongly weathered samples B-5 and B-6 also show the highest $^{87}Sr/^{86}Sr$ ratios (Fig. 2.1). This behaviour allows us to suppose that the mineral phases with low Rb/Sr and $^{87}Sr/^{86}Sr$ ratios preferentially go into solution and get transported away, as only by such a process could components with higher Rb/Sr and $^{87}Sr/^{86}Sr$ ratios remain behind.

Investigations have shown that plagioclase feldspar is most important in this respect. No study has been able to show that the Sr isotope system can be completely rehomogenized on a large scale in a soil profile. Rb-Sr isotope determinations on whole rock from a weathered soil profile are therefore unlikely to yield information about the age of an ancient weathering profile. An exception can be found in the study of Worden and Compston (1973). Worden and Compston carried out an isotopic study on the weathering profile of a granite body from Australia. The investigated samples were at various stages of weathering between fresh granite and strongly decomposed material showing quartz-kaolinite mineral paragenesis. The age of the granite was estimated at 2600 to 2700 million years. However, laterite formation was initially assigned a Cenozoic age.

As can be seen from the isochron in Fig. 2.2, the strongly weathered samples show no deviation from the isochron and, together with the fresh samples, define a reasonable age of 2580+/-16 Ma. Even the weathered material alone would give an age practically identical to the above one within margins of error. How can we interpret this isochron? The authors produce good arguments that would tend to exclude this as the effect of recent weathering.

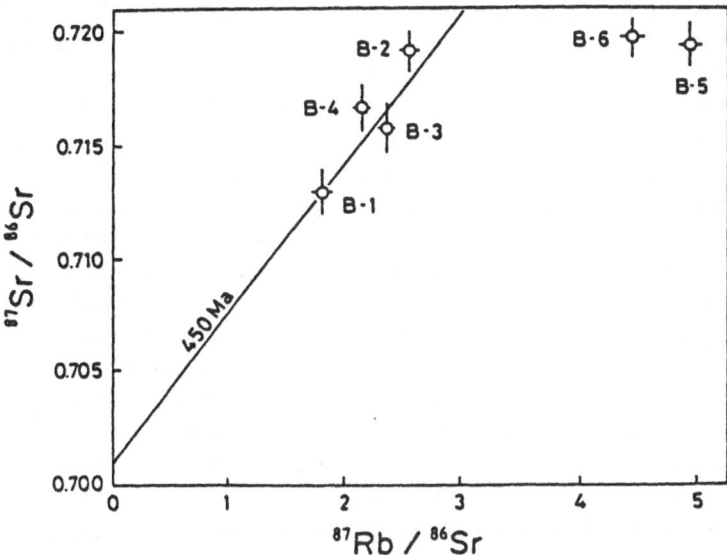

Fig. 2.1. Isochron diagram showing the influence of weathering on the Rb-Sr isotope system of samples from a 450 Ma old granite pluton. B-1: fresh rock; B-5 and B-6: strongly weathered samples. (after Dasch 1969)

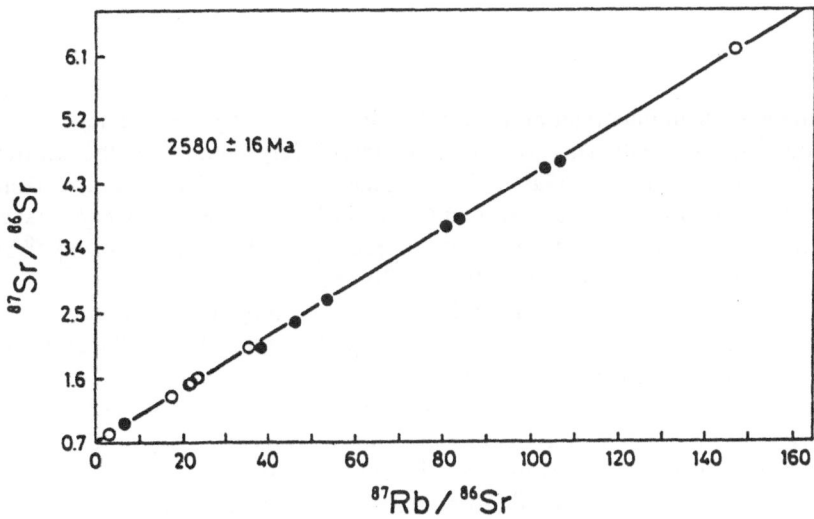

Fig. 2.2. Rb-Sr isochron diagram for fresh (open circles) and altered samples (filled circles) from a 2600 Ma old granite body. (after Worden and Compston 1973)

The fact that the weathered samples also lie on the isochron would suggest in the case of recent weathering that radiogenic ^{87}Sr and Rb of particular relative amounts had been either lost from or added to the system. The sample points would, in this case, move along the isochron during the course of weathering. Although such a process is conceivable, it does appear rather improbable. In the case of an ancient phase of weathering, it would have to be assumed that it followed closely after intrusion, as a consequence of hydrothermal activity or surface weathering, and after rapid excavation of the pluton. In such a case, the mineral's isotopic systems would not yet have developed in significantly different ways, and in the case of a pluton where the system has remained a closed system, the weathering products would have taken on a similar Sr isotopic composition.

In an isochron diagram, which displays the data at the time of the intrusion or shortly after, all the data points would plot on a near horizontal line, the exact location on the line depending on the Rb/Sr ratio. Such a model implies, however, that the weathering profile has remained a closed system since its origination. In this case, the weathering would be datable.

Let us delve a little deeper into the processes of weathering and discuss the behaviour of isotopic systems of single mineral components during weathering and during the formation of, for example, clay minerals.

2.2 Weathering of Biotite and Muscovite Mica

As biotite is such an important mineral in Rb-Sr and K-Ar age determinations, its behaviour during weathering was already intensively studied in the sixties (Zartman 1964; Goldich and Gast 1966). A general conclusion to be drawn from all these studies was that the decrease in age was a result of the relatively weaker bonds in the crystal lattice formed by the radiogenic isotopes ^{87}Sr and ^{40}Ar compared with those of the parent isotopes ^{87}Rb and ^{40}K.

The weathering process significantly controls the migration of pore waters through the rock. In a combined mineralogic and isotope study, Clauer et al. (1982) were able to observe and quantify the natural, progressive weathering of biotite in the soil profile lying above a migmatite. Samples were collected from variously weathered basement rocks as well as the soils themselves. Biotite and its weathering product kaolinite were investigated for their Rb-Sr, K-Ar, O and H isotopes. Mineralogic studies were carried out as an additional control.

The early stages of biotite weathering is marked by oxidation of Fe^{2+}, transport of K, Ti, Mn and Rb, and a strong increase in water and calcium contents. Radiogenic ^{87}Sr and ^{40}Ar were both quickly dissolved out of the biotite particles, displaying very similar mobility behavior. This can be seen in Fig. 2.3 in which both [87]Sr and [40]Ar contents are found to correlate with each other.

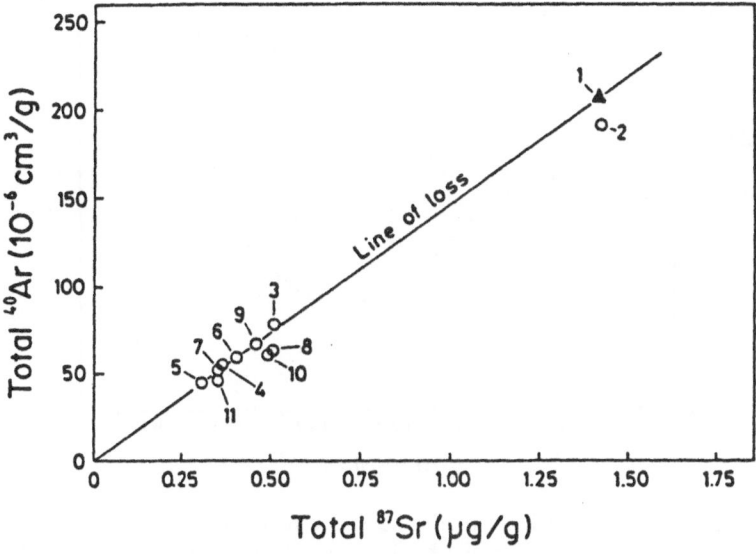

Fig. 2.3. Comparison of radiogenic [87]Sr and [40]Ar in biotites with increasing weathering. 1 and 2: fresh biotites; 3-11: altered biotites.(Clauer et al. 1982)

Table 2.1. Rb-Sr isotope data of biotites from a weathering profile. (from Clauer 1981)

sample	Rb	Sr	$^{87}Rb/^{86}Sr$	$^{87}Sr/^{86}Sr$
biotite 1	579	16.6	101.3	1.5890
biotite 1'	655	22.0	86.33	1.3730
biotite 2	544	38.8	40.66	0.8477
biotite 3	593	33.3	51.67	0.8236
biotite 4	515	60.6	24.61	0.7648
biotite 4'	526	59.0	25.83	0.7764
biotite 4"	543	60.5	26.03	0.7792
biotite 5	572	72.2	22.96	0.7836
biotite 5'	553	55.7	28.76	0.7967
biotite 6	479	92.2	15.06	0.7667
biotite 6'	386	68.0	16.43	0.7649

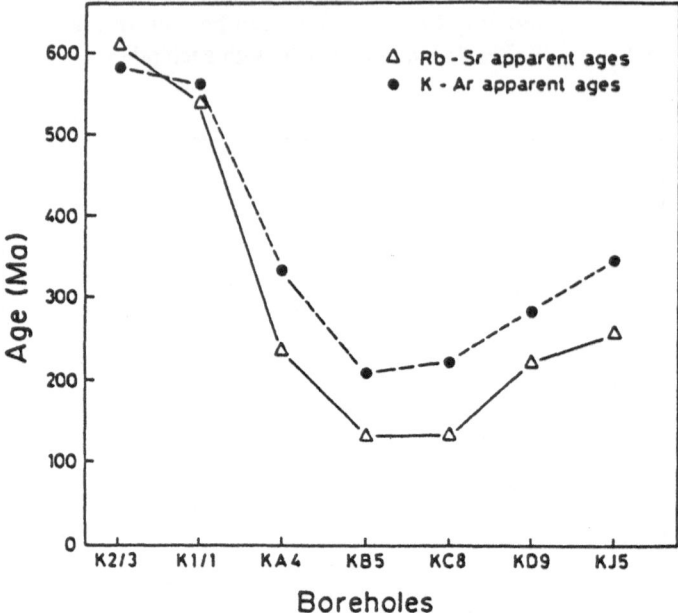

Fig. 2.4. Apparent Rb-Sr and K-Ar biotite ages as a function of increasing weathering. Fresh biotites from boreholes K2/3 and K1/1. Altered biotites from boreholes KA4-KJ5. (Clauer et al. 1982)

Table 2.2. $\delta^{18}O$ and δD values (SMOW) in altered biotites. (from Clauer et al. 1982)

sample	$\delta^{18}O$	δD	H_2O (%)
1	5.7	-65	2.77
3	7.2	-83	5.73
4	7.4	-92	5.05
5	8.5	-84	5.30
6	7.6	-84	5.29
7	8.4	-87	5.17
8	7.5	-92	4.67
9	8.0	-91	4.75
10	8.5	-87	7.14

Weathered biotites fall without exception on a straight line, which can be extended to reach the fresh biotites. This means nothing other than that Sr and Ar have been lost to the system at comparable rates. The K-Ar age produces an "apparent age" that is likely to be similar to the Rb-Sr "age" (Fig. 2.4).

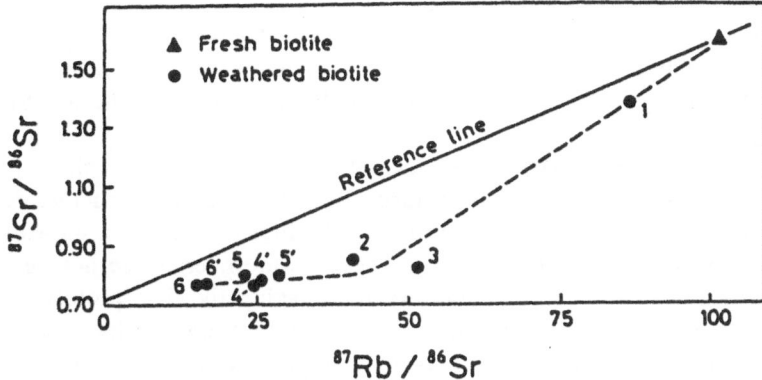

Fig. 2.5. Fresh and altered biotites in a Rb-Sr isochron diagram. The most altered biotites (4-6) appear to be near isotopic equilibrium and show much lower ^{87}Rb/^{86}Sr and ^{87}Sr/^{86}Sr ratios than fresh biotites. (Clauer 1981)

Table 2.1 allows us to establish that the Sr content of biotitesincreases with increasing weathering, whereas the ^{87}Sr/^{86}Sr and Rb/Sr ratios decrease (#1: fresh biotite; #6: most altered biotite). Therefore, radiogenic Sr dissolves and is lost from the biotite while non-radiogenic Sr is built into the lattice. This has the consequence that the sample points under the reference line move towards lower ^{87}Sr/^{86}Sr and Rb/Sr ratios (Fig. 2.5).

The δ^{18}O and δD values of 5.7‰ and -65‰, repectively, for biotite #1, are typical for unweathered biotites from magmatic rocks. This is confirmed by the low water content of 2.8%. The strong change in isotopic composition which takes place from these fresh biotites to the slightly weathered biotites (#3 and #4) is striking. The δ^{18}O values for these samples show strong enrichment in ^{18}O, with δ^{18}O rising almost 2 per mil, from 5.7‰ to 7.4‰. Despite obvious signs of weathering, further increase is restricted to just 1.4‰. The initial climb in δD values is even more marked. The values jump by around 20-30‰ and remain constant thereafter, apparently oblivious to further increases in weathering. This may indicate that even the further chemical weathering of previous biotites has no effect on the relative deuterium content. Coexistent with the extreme changes in the isotopic compositions is an increase in the amount of water, which almost doubles relative to fresh biotite. This sudden switch, not only in isotopic composition, but also in water content, hardly speaks for a gradual biotite weathering process.

An interesting aspect of these sudden changes in isotopic composition is that they are not accompanied by any visible or measureable new formation of weathering product minerals. Neither microscopic or X-ray investigations allow for recognition of new minerals in samples #3 and #4. Yet it can be assumed that even in these early stages of weathering, crystallographic changes to the mineral structure have begun to take place as only such changes would allow for isotopic exchange under low temperature conditions. It is still left to further research in mineralogy and crystallography to tackle and solve this particular problem. New formation of clay minerals in the form of kaolinite can only be recognized in sample #5. During this stage, biotite shows only minor variation in its $^{87}Sr/^{86}Sr$ ratio (Fig. 2.6).

On the basis of these observations, the authors suggest that amorphous proto-minerals or proto-clay minerals have formed inside the biotite structure during weathering, a process that is recorded in the significant changes in the isotopic compositions of oxygen and hydrogen. These proto-minerals are not yet well enough developed to be identifiable crystallographically. The prototypes are transitory in the genesis of clay minerals and represent more or less metastable phases. The more stable authigenic clay mineral phases can only start developing between mineral particles and the environment once chemical and isotopic equilibrium has been reached. Muscovite is quite resistant against weathering compared with biotite and is not an important precursor of clay minerals in the weathering milieu.

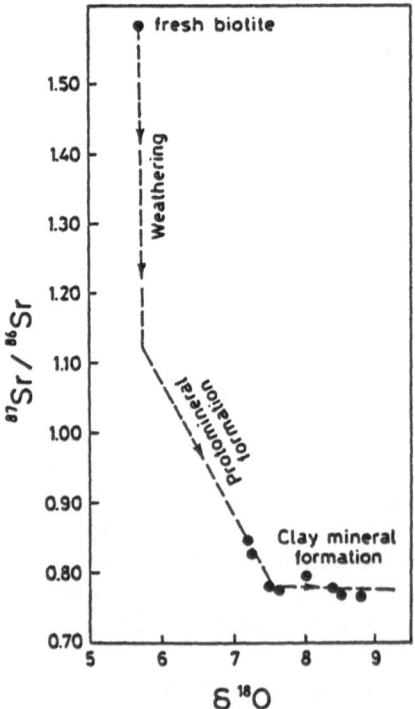

Fig. 2.6. Evolution of $^{87}Sr/^{86}Sr$ and $\delta^{18}O$ during alteration of biotites and resulting new formation of clay minerals. (Clauer and Chaudhuri 1995)

Muscovite can, however, be broken down through mechanical erosion and transport to attain clay mineral grain sizes. Investigations carried out by Clauer (1981) on muscovites from a partly fresh, partly strongly weathered pegmatite of the Congo region demonstrated that the K-Ar and Rb-Sr ages had not been changed during the weathering process. He inferred that the isotopic systems would not be disturbed as long as the crystallographic characteristics did not change.

2.3 Weathering of Feldspars

Several clay minerals such as illite, kaolinite and gibbsite can be produced by the continual weathering of feldspars. However, how their isotopic systems behave has been only scantily studied. Observations made by Goldich (1938) allowed him to assume that feldspars, like biotite, were particularly sensitive to weathering. Only in tropical soils are the Rb-Sr and K-Ar isotopic systems in feldspars more resistent to weathering than those of biotite (Zartman 1964; Clauer 1981).

Rb-Sr isotope data from feldspars investigated by Clauer (1981) from a weathering profile in the Chad republic are displayed in Fig. 2.7A. The gradient of the reference line corresponds to the age of intrusion. With the exception of one sample, the feldpars lie near or on the reference line. They appear therefore to have remained stable relative to weathering. This is also the case for the K-Ar system of the feldspars (Fig. 2.7B). The reference line in this case corresponds to to the age of intrusion, too.

In the Sr diagram, one sample #F5 displays higher $^{87}Sr/^{86}Sr$ ratios relative to the other samples (Fig. 2.7A). This sample was collected at great depth within the weathering profile. Several investigations have shown that feldspars taken from close to the surface in tropical soils are far less weathered than those from greater depths. Close to the surface, the insides of the feldpars remain relatively well preserved, protected by a thin clay mineral coating on the crystal surfaces. Deeper in the weathering profile, the cores of the feldspars are altered with significant transformation into various clay minerals. This can also be observed in sample #F5. Mechanical action on the sample whilst in the ultrasonic bath has the consequence that the feldspars from close to the surface are freed from their clay mineral coating. These near surface samples help to define the magmatic age.

The strongly transformed feldspars from depth, enriched in clay minerals, contain a high $^{87}Sr/^{86}Sr$ ratio which does not represent the magmatic isotopic composition any more but reflects more the influence of the surrounding weathering milieu. It is possible that radiogenic Sr was released during the weathering of biotite and has been incorporated into the clay minerals.

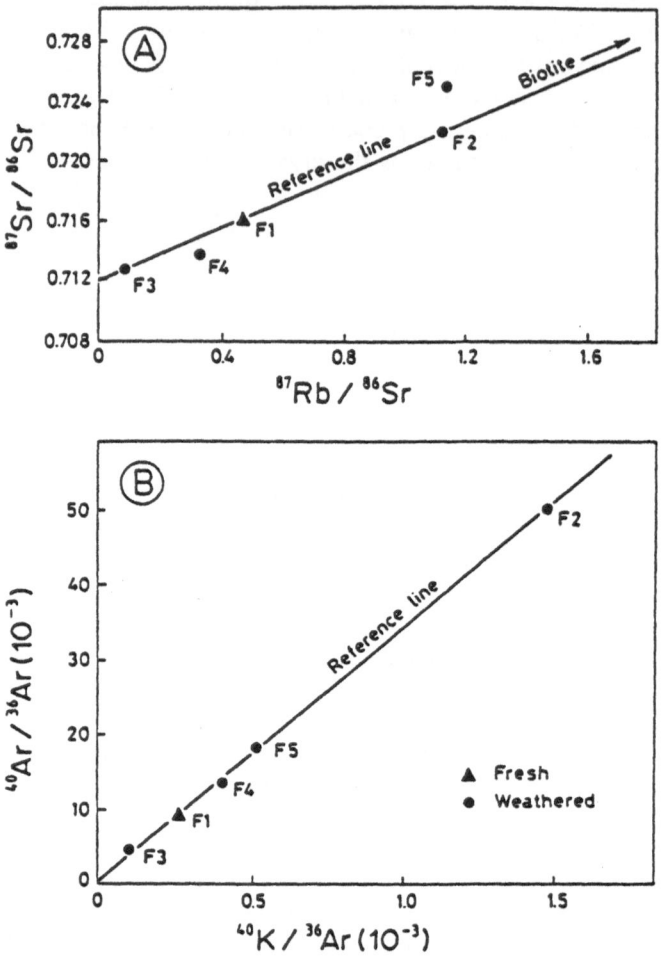

Fig. 2.7. Rb-Sr (A) and K-Ar (B) isochron diagrams for fresh and altered feldspars. (Clauer 1981)

2.4 Products of Chemical Weathering

Most of the important clay minerals are derived from biotites and feldspars. They determine the Rb-Sr and K-Ar isotope characteristics of the weathering profile. Let us turn to these products of weathering themselves, their isotopic systems and behavior during weathering.

Clauer (1979) investigated exchange reactions that took place between kaolinite and fluid phases within a weathering profile. The research was carried out on the weathering profile of a pegmatitic dyke, over 2.6 Ga old.

Kaolinite was collected from the very top of the profile, 8 m lower near the groundwater table and below the groundwater table near the base of the profile. The Rb-Sr concentrations and the $^{87}Sr/^{86}Sr$ ratios were strongly dependent on the surrounding milieu and level in the weathering profile (Table 2.3). Low Sr contents of 0.9-7.9 ppm (K5, K6) and high $^{87}Sr/^{86}Sr$ ratios between 0.980 and 2.313 were found from kaolinite uppermost in the profile. Around the groundwater table, the Sr concentration rises to 120 ppm (K4) and the $^{87}Sr/^{86}Sr$ ratio falls to 0.7604. Below the groundwater table, the Sr concentrations decrease to 50 ppm and $^{87}Sr/^{86}Sr$ ratios increase to 0.827 (K3).

Table 2.3. Sr isotope data of kaolinites. (from Clauer 1979)

sample	Rb (ppm)	Sr (ppm)	Rb/Sr	$^{87}Sr/^{86}Sr$
K3	46.9	49.2	0.95	0.82700
K4	60.2	122.4	0.49	0.76042
K5	47.6	7.9	6.03	0.9803
K6	23.3	0.91	25.60	2.3128

The kaolinites define a Rb-Sr isochron age of 1770+/- 30 million years. Is this age geologically significant? Does it reflect the age of Sr homogenization in the profile? The Sr mixing diagram (Fig. 2.8; see Faure 1986) allows us to assume that this age is of no geologic significance. The sample points in this diagram depict a straight line and demonstrate that the compositions of these kaolinites have been influenced by two very different sources, at least with respect to strontium.

Clauer (1979) assumes that the Sr in sample K6 was incorporated during the crystallization of the kaolinite. In his opinion, the components with high Sr contents and a $^{87}Sr/^{86}Sr$ ratio of 0.78 reflect the chemistry of the groundwater and the circulating surface water. This strontium was possibly not built into the crystal lattice but instead adsorbed onto crystal surfaces. Investigations carried out by Fordham (1973) show clearly that Sr preferentially takes up adsorption sites on kaolinite crystal surfaces. Thus, kaolinite captures external, exchangeable strontium.

Clauer's study (1979) demonstrates that weathering exchange and mixing processes are hardly even centimeter-scale phenomena. Thermodynamic investigations by Garrels and Christ (1965), Fritz and Tardy (1973) and Tardy and Garrels (1974) clearly show that minerals in weathering profiles strive for chemical equilibrium with the surrounding fluid phases, but that such equilibrium and in particular isotopic equilibrium can only be achieved in the smallest micro-environment.

In order to understand chemical exchange processes during weathering better, hydrogen and oxygen isotope investigations in particular were carried out on clay minerals. We now know that chemical reactions that eventually lead to the formation of clay minerals can only take place under conditions of high water/rock ratio. This means, that a large amount of water must exchange with rock in order for a clay mineral to be formed. Based on hydrogen and oxygen isotope studies, it can be safely assumed that such authigenic clay minerals exist in isotopic equilibrium with local, meteoric water.

As water is the most important medium in which the weathering of minerals and the new formation of clay minerals takes place, we need to take a look at the isotopic characteristics of water and its isotopic fractionation processes.

When water evaporates at the surface of the ocean, the water vapour is enriched in ^{16}O and H because $H_2^{16}O$ has a higher vapour pressure than $H_2^{18}O$ or deuterium water molecules (Fig. 2.9). As a consequence, the $\delta^{18}O$ and δD values of water vapour above the ocean are negative with respect to "Standard Mean Ocean Water" or "SMOW". If rain drops form as a result of condensation of the water vapour in a cloud, these rain drops will be relatively enriched in ^{18}O and D, as is seawater. This removal of ^{18}O and D has the consequence that the remaining damp air masses become ever more enriched in ^{16}O and H.

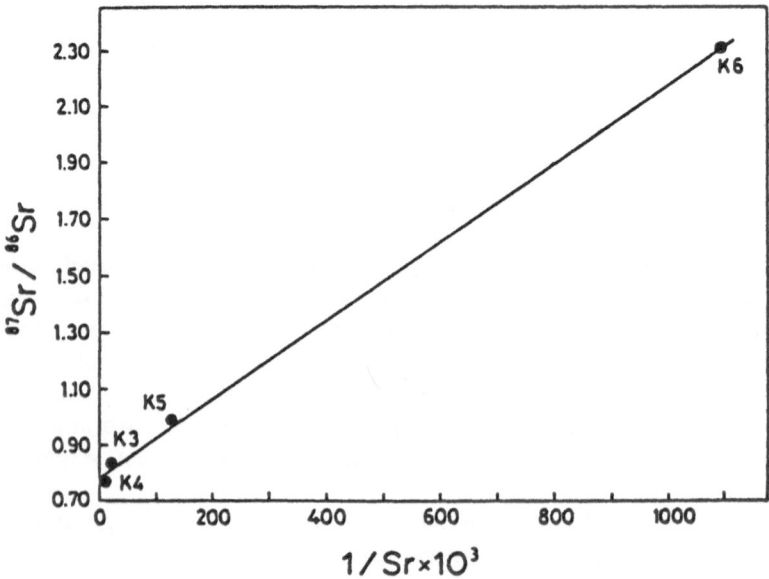

Fig. 2.8. Sr mixing diagram for kaolinites from a weathering profile. (Clauer 1979)

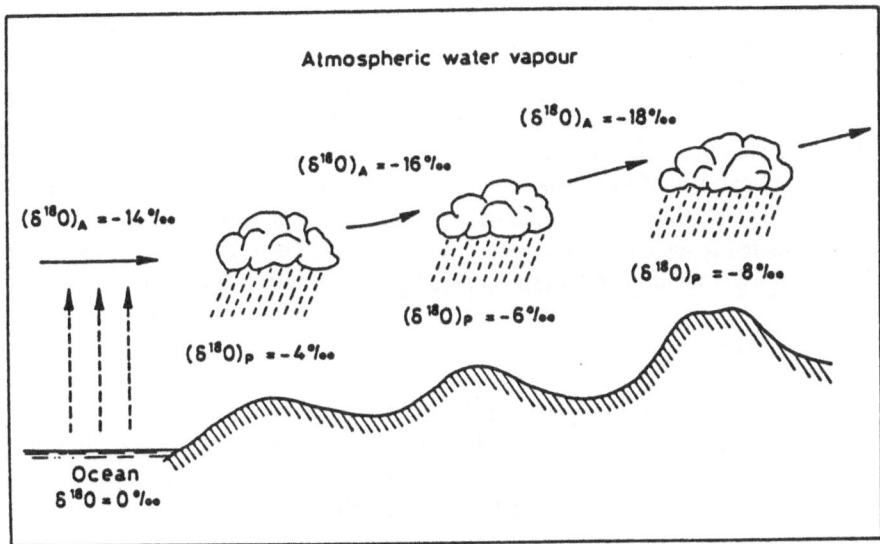

Fig. 2.9. Oxygen isotope fractionation during evaporation and condensation. $(\delta^{18}O)_A$: isotopic composition of atmospheric water vapour: $(\delta^{18}O)_P$: isotopic composition of precipitation. (modified after Welhan 1987)

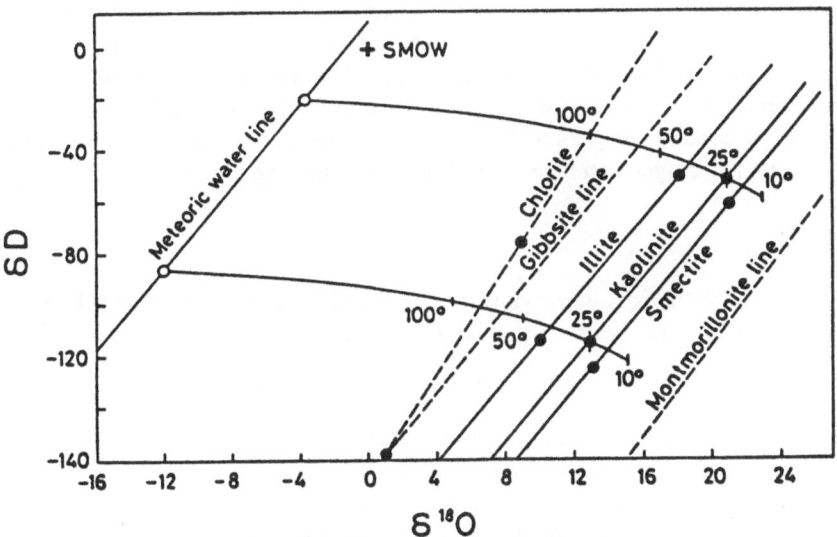

Fig. 2.10. δD and $\delta^{18}O$ values of hydrous minerals formed in equilibrium with meteoric waters. (Kyser 1987)

The more precipitation that takes place, in the form of rain, snow or hail, the more negative will be the $\delta^{18}O$ and δD values of the remaining water vapour. This process leads to significant isotopic fractionation. As a result, fresh water is generally very poor in ^{18}O and D. Conversely, fresh water is relatively enriched in the lighter isotopes, ^{16}O and H, with respect to seawater. With the help of a great number of analyses of meteoric waters, collected at various places on the Earth, Craig (1961) was able to show that the $\delta^{18}O$ and δD values of meteoric waters are linearly related to each other ("meteoric water line"). The relationship between the $\delta^{18}O$ and the δD values can be represented by the following equation:

$$\delta D = 8\delta^{18}O + 10$$

Let us return to the clay minerals. If the clay minerals were in isotopic equilibrium with meteoric waters and formed at the same temperature, then their $\delta^{18}O$ and δD values should come to lie on a straight line, which is parallel to the meteoric water line (see equation III, Sect. 1.1). Experiments have shown that this is indeed the case. 25°C isotherms or weathering lines for illite, kaolinite, smectite and montmorillonite are shown in a $\delta^{18}O$ and δD diagram (Fig. 2.10). A clay mineral may form at different temperatures and incorporates water with various $\delta^{18}O$ and δD values depending on the temperature of formation. Fig. 2.10 shows kaolinite which formed at 10, 25, 50 and 100°C.

A study concerning isotopic equilibria in soil was carried out by Lawrence and Taylor (1971). The authors determined the oxygen and hydrogen isotopic compositions of clay minerals from 75 Quaternary soils and associated waters of various regions of the USA (Fig. 2.11). The $\delta^{18}O$ and δD values for meteoric waters confirm the Craig "meteoric water line". The observed differences in the isotopic compositions of clay minerals and meteoric waters approach those of calculated fractionation factors and allow us to assume that isotopic equilibrium for oxygen and hydrogen can be achieved in the weathering milieu.

2.5 The Sr Isotopic Composition of Fluid Phases in Weathered Profiles; an Isotopic Model for Rock Weathering

As we have seen, groundwater, circulating surface water and pore water influence both the course of weathering and the processes that lead to the formation of clay minerals. Only by the exchange of a large quantity of water with rock can clay minerals form, their oxygen and hydrogen isotopic compositions being in isotopic equilibrium with the circulating fluid phases. The information provided by the preceding discussion allows us to present an isotopic model of rock weathering that may be comparable with nature. This model will be important in understanding the behavior of Sr during diagenetic processes as similar mechanisms are also involved during the sedimentation and diagenesis of sediments.

Of the Sr-rich minerals, apatite and the feldspars (particularly plagioclase) weather more rapidly than biotite and muscovite, making it possible to reconstruct the ideal, near consecutive weathering of a rock.

Let us start with the Rb-Sr mineral isochron of a granite. The minerals should lie on one isochron, providing they all crystallized out at the same time from the same magma, having incorporated the same initial $^{87}Sr/^{86}Sr$ ratio as the magma body (Fig. 2.12A). Due to its high Sr and low Rb content, plagioclase (Pl) lies to the left of the whole rock (WR), whereas alkali feldspar (Kf) and biotite (B) with their relatively low Sr contents and high Rb contents lie to the right of WR. The position of whole rock, which must lie between the two end-member mineral components, arises from the law of mass balance.

Every chemical weathering process can be compared to a leaching experiment whereby chemical elements from various minerals of the bulk rock are sequentially leached. For the sake of simplicity, we assume that our whole rock has remained a closed system during weathering and no element has been introduced from outside. We also assume that weathering attacks only the plagioclase first.

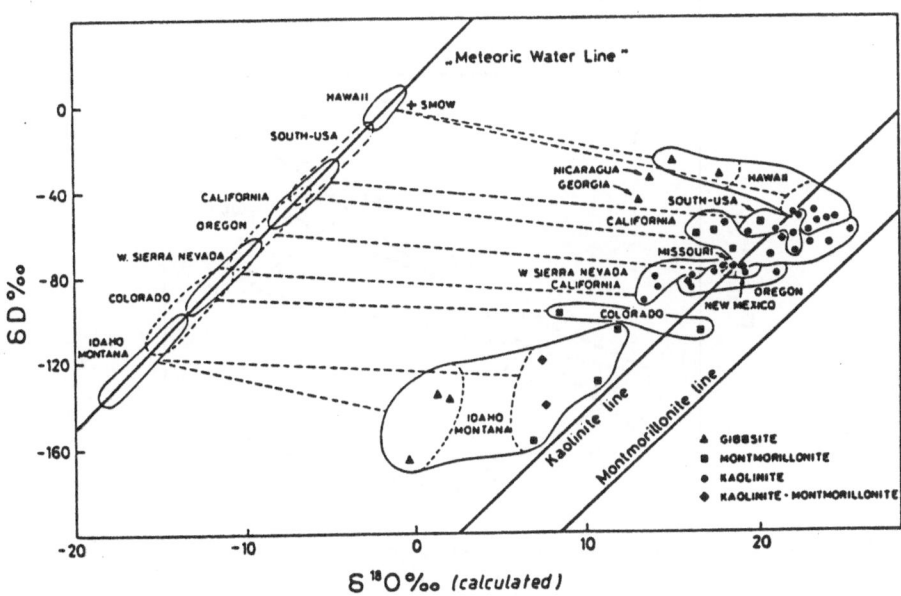

Fig. 2.11. δD and δ^{18}O values of gibbsites, montmorillonites, kaolinites in soil profiles in North America and Hawaiian Islands to corresponding local present-day meteoric waters. (Lawrence and Taylor 1971)

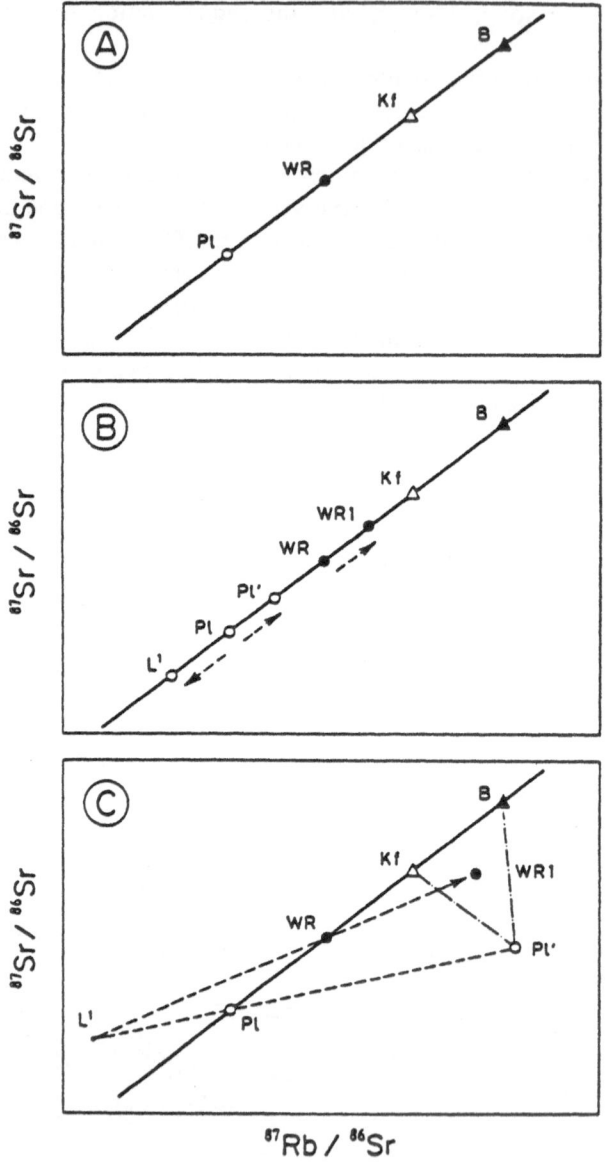

Fig. 2.12. Sr isotope model of rock weathering. A: Rb-Sr isochron for plagioclase (Pl), whole rock (WR), biotite (B) and alkali feldspar (Kf). B: Leaching of plagioclase (Ll=leachate; Pl'=plagioclase residue) causes an increase of the Rb/Sr and $^{87}Sr/^{86}Sr$ ratios in the whole rock (WR->WR1). C: Selective leaching of Sr out of Ca- and Sr-rich zones of the plagioclase (Rb remaining immobile) causes an increase of the Rb/Sr ratio in the residual plagioclase (Pl') and consequently in the whole rock (WR1). The pore water Ll, which exchanged with the plagioclase has high Sr and low Rb contents and moves away from the isochron towards lower Rb/Sr ratios.

This has the consequence that the fluid phases mainly dissolve Sr with relatively low ^{87}Sr/^{86}Sr and Rb/Sr ratios, which goes on towards defining pore water chemistry. The pore waters will be strongly enriched in Sr. The whole rock does not lose much radiogenic Sr and so, its ^{87}Sr/^{86}Sr and Rb/Sr ratios must rise. If Rb and ^{87}Sr were not fractionated by this weathering process, Sr would not dissolve preferentially relative to Rb, and the whole rock position would move up the isochron towards even higher ^{87}Sr/^{86}Sr and Rb/Sr ratios (WR1; Fig. 2.12B).

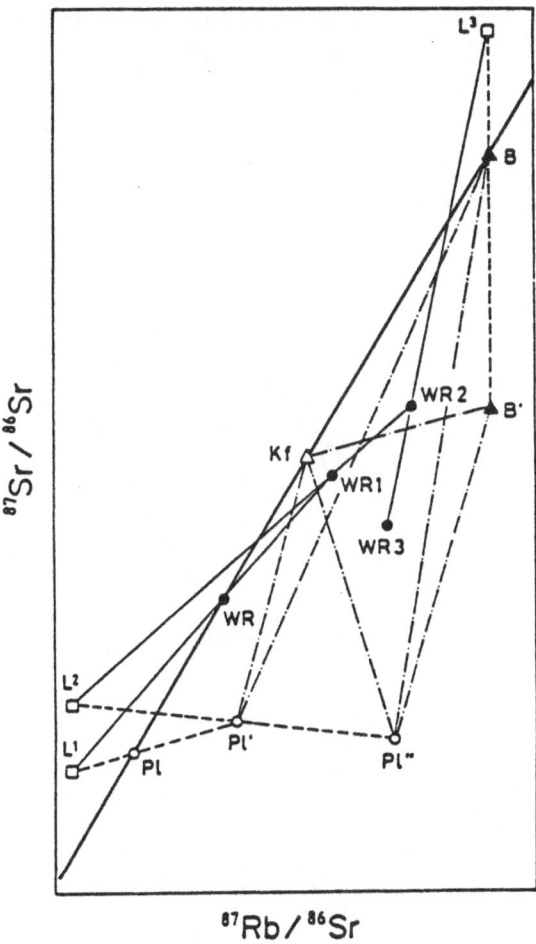

Fig. 2.13. Sr isotope model of rock weathering. With the onset of weathering of biotite and alkali feldspar the isotope evolution of Sr in the whole rock begins to change (WR1->WR2 ->WR3). Mass balancing tells us that the isotopic composition of the altered whole rocks WR1,WR2 and WR3 must lie within the triangles Kf-Pl'-B, Kf-Pl''-B and Kf-Pl''-B', respectively.

However, this is a special case and is seldom observed in nature. The different physico-chemical characteristics of Rb and Sr, their incorporation into different lattice sites and chemical inhomogeneities in the crystal lattice all induce fractionation of Rb and Sr. Studies show that Sr is dissolved from the Ca-Sr rich zones of a plagioclase during early stages of weathering. If Sr is preferentially dissolved and Rb remains relatively immobile, the whole rock position will move away from the isochron towards higher Rb/Sr ratios. Fig. 2.12C sheds light on the isotopic evolution of Sr during this stage of weathering. Selective leaching may leave a residue Pl' and a fluid phase L1. This allows us to reconstruct the isotopic composition of the weathered rock. If we assume that the rock consists of three phases, plagioclase residue, biotite and alkali feldspar, then, mass balancing tells us that the isotopic composition of the whole weathered rock (WR1) must lie within this triangle.

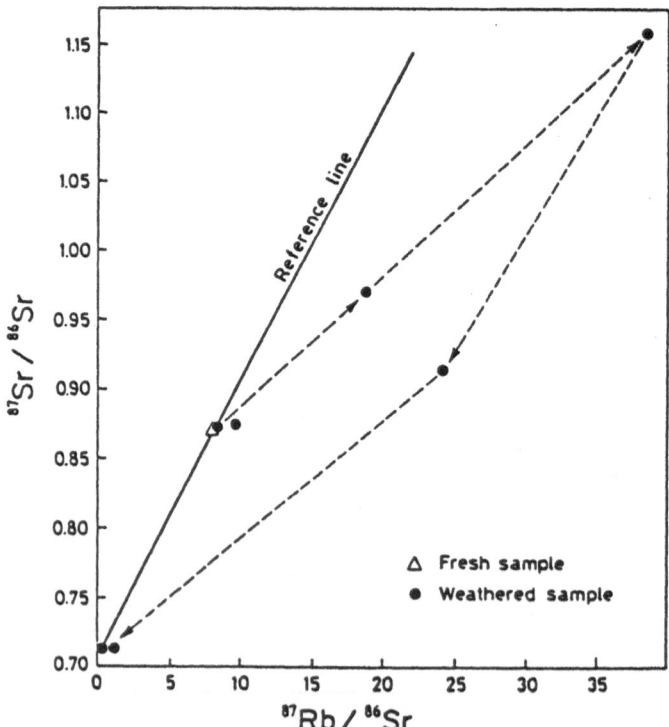

Fig. 2.14. Rb-Sr isochron diagram for fresh and altered whole rock samples from the Butler Hill granite. Samples, which are not altered, plot on the reference line defining age and initial ratio of the intrusive complex. The arrows indicate increasing alteration. (after Blaxland 1974)

On the other hand, the assumption that the system remains closed throughout means that pore water (L1) and both fresh and weathered rock (WR, WR1 respectively) must all lie on one straight line. Continued weathering would lead to the isotopic evolution (WR-->WR1-->WR2-->WR3; Fig. 2.13). Only with the onset of weathering of biotite and alkali feldspar does the evolution of Sr in the whole rock begin to change. Two effects start to influence the isotopic composition significantly. First, biotite releases radiogenic ^{87}Sr (Sect. 2.2), lowering its ^{87}Sr/^{86}Sr ratio. Second, biotite will tend to adsorb more and more Sr from the surrounding pore waters and perhaps lower its Sr concentration. This pore water will be strongly enriched in only weakly radiogenic Sr due to the previous weathering of plagioclase (Sect. 2.2). This has the consequence that the ^{87}Sr/^{86}Sr and also the Rb/Sr ratios of the whole rock decrease (evolution of WR2 into WR3; Fig. 2.13). The model helps to explain the positions of the sample points for the Elberton granite in the isochron diagram (Fig. 2.1) or those observed from other weathering profiles, e.g. the Butler Hill granite (Blaxland 1974; Fig. 2.14). The course of weathering outlined here results in certain rules of thumb concerning the isotope geochemistry of pore waters during weathering (and diagenesis). Pore waters in the early plagioclase weathering stages are likely to show high Sr contents and less radiogenic isotopic signatures, whereas those of later stage alkali feldspar and biotite weathering allow the development of porewaters with lower Sr contents (due to Sr adsorption of biotites) and strongly radiogenic isotope characteristics.

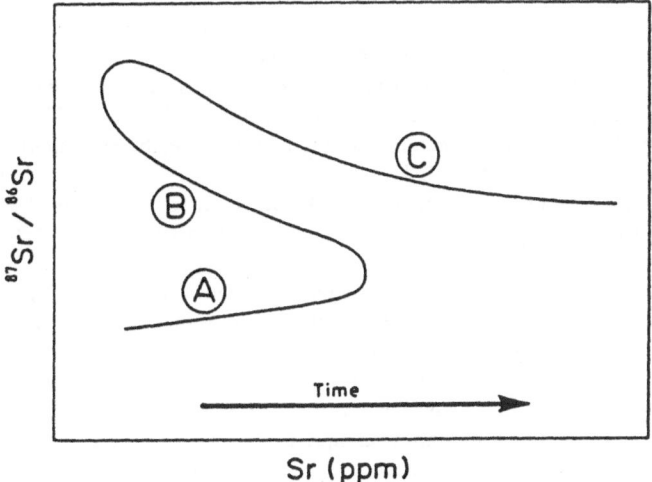

Fig. 2.15. Schematic representation of the Sr isotopic evolution of pore waters during weathering of a granite rich in plagioclase. A: alteration of plagioclase; B: increasing alteration of biotite and alkali feldspar; C: biotite and alkali feldspar in chemical

equilibrium with the surrounding porewaters and advanced weathering and dissolution of plagioclase.

Strictly speaking, the process of weathering individual minerals does not proceed stepwise but instead with considerable overlap. The isotopic compositions and the Sr concentrations of pore waters are products of the mixing of each individual fluid phase that has come about by the weathering of previous mineral components. The course of curve A (Fig. 2.15) represents pore water deriving from the weathering of plagioclase. Increasing alkali feldspar and biotite weathering results in curve B with increasing $^{87}Sr/^{86}Sr$ ratios and decreasing Sr concentrations. Thermodynamic calculations demonstrate that curve C only comes into play when chemical equilibrium has been reached between biotite, alkali feldspar and the surrounding fluid phases (Probst et al. 1992). Advanced weathering and dissolution of plagioclase leads to decreasing $^{87}Sr/^{86}Sr$ ratios and increasing Sr concentrations. Thermodynamic calculations show clearly that the course of weathering depicted in Fig. 2.15 is greatly accelerated today by the presence of acid rain (Probst et al. 1992). Clay minerals, which crystallize out in the fluid phase, are likely to display highly variable isotopic compositions depending on when they form.

2.6 References

Blaxland AB (1974) Geochemistry and geochronology of chemical weathering, Butler Hill granite, Missouri. Geochim Cosmochim Acta, 38:843-852

Clauer N (1979) Relationship between the isotopic composition of strontium in newly formed continental clay minerals and their source material. Chem Geol, 27:115-124

Clauer N (1981) Strontium and Argon isotopes in naturally weathered biotites, muscovites and feldspar. Chem Geol, 31:325-334

Clauer N O'Neil JR, Bonnot-Courtois C (1982) The effect of natural weathering on the chemical and isotopic compositions of biotites. Geochim Cosmochim Acta, 46:1755-1762

Craig H (1961) Isotopic variations in meteoric waters. Science, 133: 1702-1703

Dasch EJ (1969) Strontium isotopes in weathering profiles, deep sea sediments and sedimentary rocks. Geochim Cosmochim Acta, 33: 1521-1552

Fordham AW (1973) The location of iron-55, strontium-86 and iodide-125 sorbed by kaolinite and dickite particles. Clays Clay Min 21: 175-184

Fritz B, Tardy Y (1973) Etude thermodynamique du systeme gibbsite-quartz-kaolinite-gaz carbonique - Application a la genese des podzols et des bauxites. Sci Geol Bull, Strasbourg, 26: 39-367

Garrels RM, Christ CL (1965) Solutions, Minerals and Equilibria. Harper and Row, New York

Goldich SS (1938) A study in rock-weathering. J Geol, 46: 17-58

Goldich SS, Gast PW (1966) Effects of weathering on the Rb-Sr and K-Ar ages of biotite from the Morton gneiss, Minnesota. Earth Planet Sci Lett, 1:372-375

Lawrence JR, Taylor HP (1971) Deuterium and oxygen-18 correlation: clay minerals and hydroxides in Quaternary soils compared to meteoric waters. Geochim Cosmochim Acta, 35: 993-1003

Probst A, Fritz B, Stille P (1992) Consequence of acid deposition on natural weathering processes: field studies and modelling. In: Kharaka YK, Maest AS Water Rock Interaction. Balkema, Rotterdam, Brookfield, pp 581-584

Tardy Y, Garrels RM (1974) A method of estimating the Gibbs energies of formation of layer silicates. Geochim Cosmochim Acta, 38:1101-1116

Welhan AJ (1987) Stable isotope hydrology. In: Kyser TK, Short course of low temperature fluids. Miner Assoc of Canada, Saskatoon pp 129-161

Worden JM, Compston W (1973) A Rb-Sr isotopic study of weathering in the Mertondale granite, Western Australia. Geochim Cosmochim Acta, 37: 2567-2576

Zartmann RE (1964) A geochronologic study of the Lone Grove pluton from the Llano Uplift, Texas. J Petrol, 5: 359-408

3 Isotope Geochemistry of River Water

Weathering is accompanied by the transport of erosion products by wind and water. Rivers and streams are responsible for removing the larger part of these erosion products in either dissolved or particular form and, for their eventual discharge into the world's ocean basins. The isotopic composition of particles and dissolved ions in rivers is determined by the average isotopic composition of the source region. The isotopic composition of the ocean is also partly governed by the constant influx of river or wind transported material (Goldstein et al. 1984). However, knowledge of the geochemical and isotopic characteristics does not just allow us to calculate mass balances for the ocean but also to identify specific river systems as sources for toxic elements (see also Chap. 4).

In natural waters, there are numerous types of particles which are carried in suspension. These may be biologic components (algae, bacteria, pieces of shells, etc.) or inorganic compounds, such as clay minerals, oxides, hydroxides (e.g. Al_2O_3, $Fe(OH)_3$, MnO_2), phosphates and carbonates. The sizes of these particles in suspension may be different, ranging from <0.1 μm up to mm-scales (Fig. 3.1). As the transport of chemical elements can take place in either particular or dissolved form depending on the physico-chemical conditions, studies must concern themselves with the dissolved as well as the suspended phases.

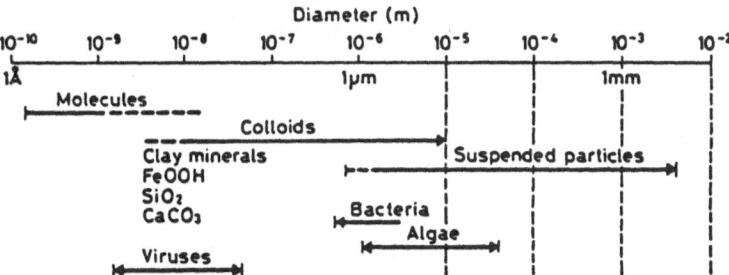

Fig. 3.1. Size comparison of biologic components and inorganic compounds. (after Sigg and Stumm 1989)

The first comprehensive, combined Rb-Sr and Sm-Nd isotope study of the dissolved and suspended loads of various rivers was carried out by Goldstein and Jacobsen (1987). They investigated major rivers in Canada, USA, Australia, Japan and the Philipines. In this study, both the Sr and the Nd isotopic compositions could be compared between the suspended load and the filtered water.

3.1 Sr Isotopes in River Water

The concentration of dissolved Sr in the various river waters studied ranged between 1 and 3000 $\mu g/l$. Sr concentrations are correlated with the major elements Na, Ca and Mg (Fig. 3.2).

The global, average (Na+Ca) concentration for river water is 18.6 mg/l. From this it is possible to work out an average river water concentration for strontium of 60 $\mu g/l$. The Sr/(Na+Ca) ratios reflect the source regions of these elements and therefore the drainage basins of the various rivers. The ratios lie understandably between those for silicate rich rocks of the upper continental crust and those for carbonate rich rocks and evaporites (1×10^{-3}; Fig. 3.2).

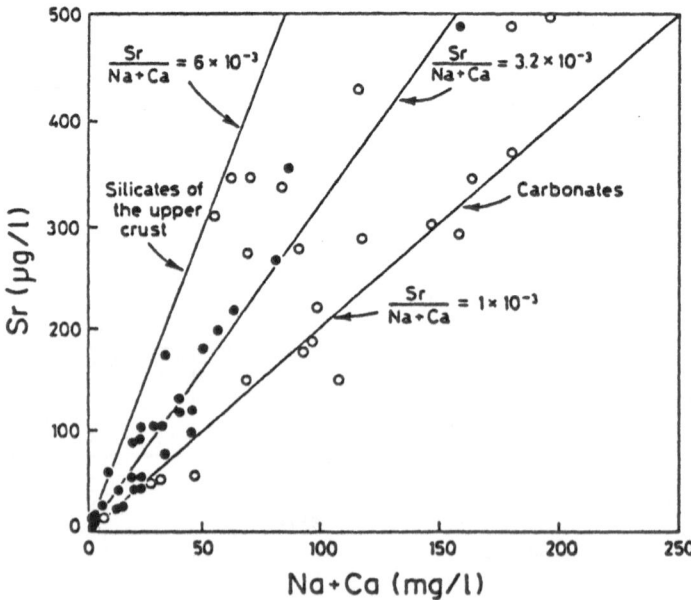

Fig. 3.2. Sr, Na and Ca concentrations of filtered river waters. Data from Goldstein and Jacobsen 1987 (filled circles) and Tricca 1997 (open circles: rivers from the upper Rhine valley)

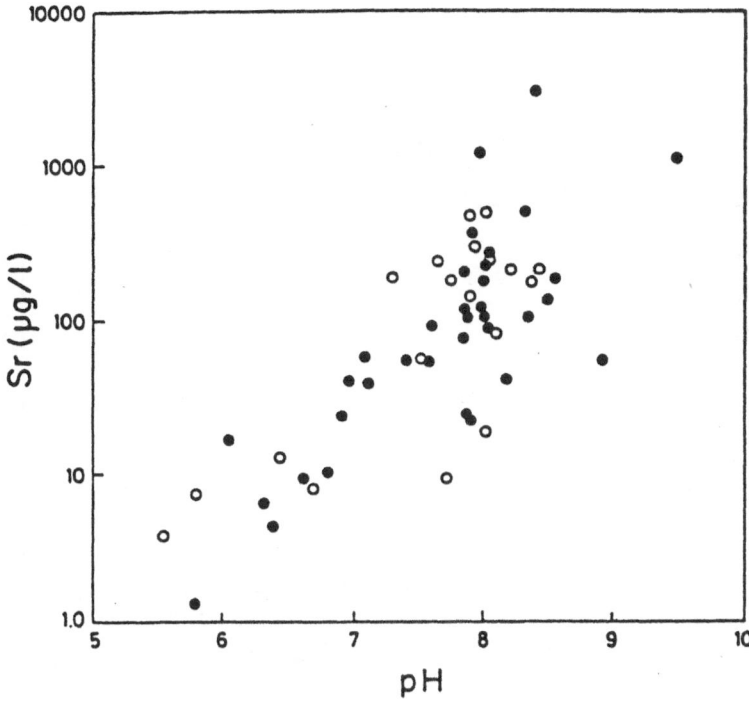

Fig. 3.3. Sr concentration of filtered river water as a function of pH. Data from Goldstein and Jacobsen 1987 (filled circles) and Tricca 1997 (open circles: rivers from the upper Rhine valley).

The Sr concentrations are strongly pH dependent (Fig. 3.3). Fig. 3.4 compares the $^{87}Sr/^{86}Sr$ ratios of filtered river water with those for the suspended load. Ratios of below 0.709 show almost identical isotopic compositions for both. Rivers whose $^{87}Sr/^{86}Sr$ ratios are lower than 0.709 come from Japan and the Philipines. Here, the drainage basins of the rivers are dominated by youthful volcanism. The young ages and low Rb/Sr ratios of the volcanic rocks prevent the development of significant variation in their $^{87}Sr/^{86}Sr$ ratios. Thus, their weathering products also show quite small variation in isotopic composition. This is the reason for the similarity between dissolved and suspended load isotopic compositions. On the other hand, rivers that drain ancient granitoid complexes show strongly variable isotopic compositions for the suspended loads and filtered waters as such source areas are marked by very variable isotopic compositions.

Radiogenic Sr is less likely to be transported into the ocean in dissolved form than in particle form. The most important carrier of strontium with high $^{87}Sr/^{86}Sr$ ratios are clay minerals. During biotite weathering, the $^{87}Sr/^{86}Sr$ ratio in the fluid phase rises (Sect. 2.2) and must help to increase the $^{87}Sr/^{86}Sr$ ratio of river water.

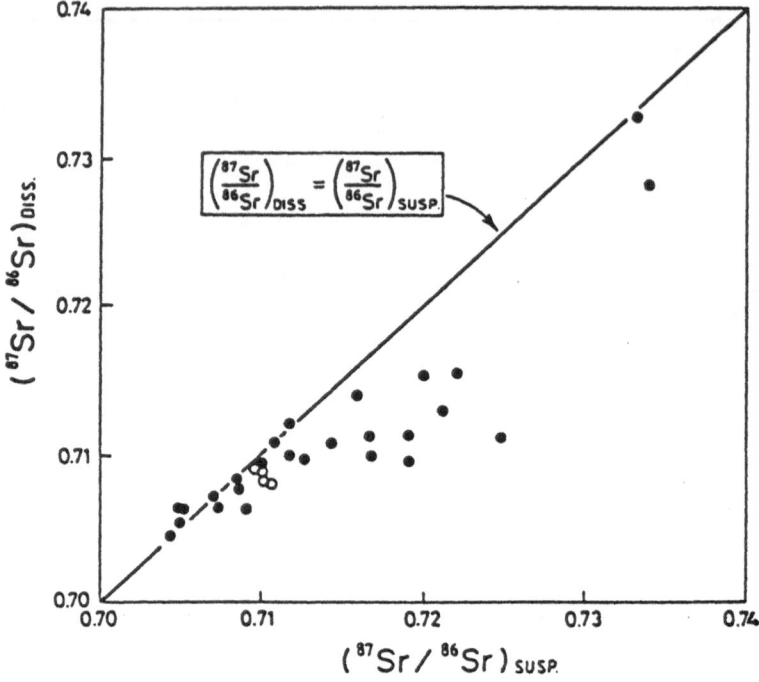

Fig. 3.4. Relationship between $^{87}Sr/^{86}Sr$ of dissolved and suspended load. Data from Goldstein and Jacobsen 1987 (filled circles) and Tricca 1997 (open circles: rivers from the upper Rhine valley)

During feldspar weathering, Sr with low $^{87}Sr/^{86}Sr$ ratios are preferentially dissolved away first. This may also have a significant effect on the isotopic composition of river water. Mass balance calculations demonstrate that the dissolution of carbonates, evaporite minerals and phosphate with low $^{87}Sr/^{86}Sr$ ratios (<0.71) has likewise a decisive influence on the Sr budget of river systems (e.g. Goldstein and Jacobsen 1987).

3.2 Nd Isotopes and the Rare Earths (REE) in River Water

The Nd and Sm concentrations in rivers vary between about 3 and 3000 ng/l, and 1 and 900 ng/l, respectively. Detailed investigations along individual river systems clearly demonstrate that they resemble Sr concentrations in that they are strongly pH dependent (Fig. 3.5; Elderfield et al. 1990). Rivers with pH values >7.5 show concentrations between 3 and 50 ng/l, whereas rivers with a pH of <7.5 display increasingly high Nd concentrations.

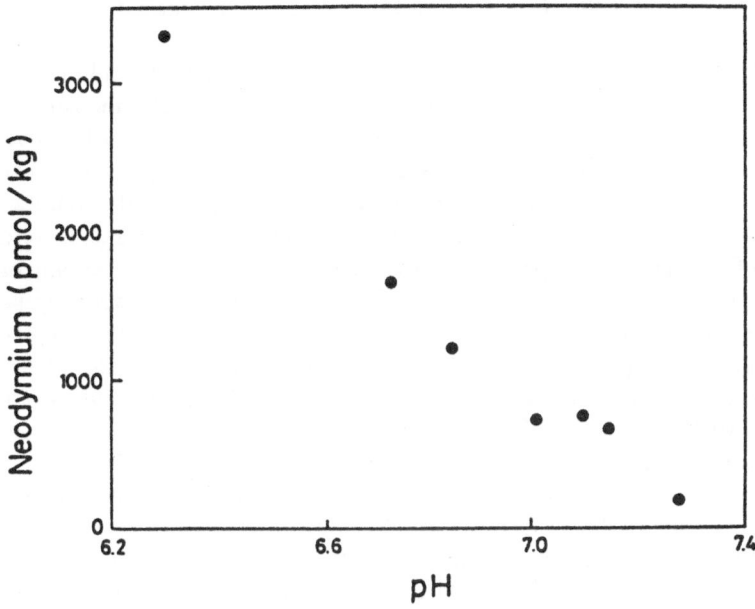

Fig. 3.5. Nd concentration of filtered river water as a function of pH. (Elderfield et al. 1990)

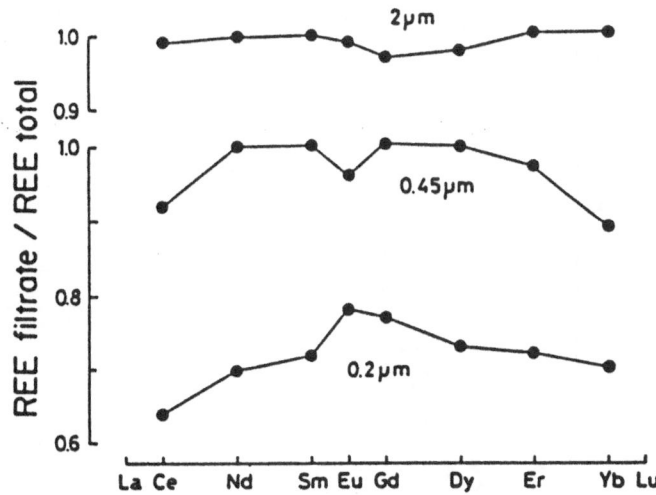

Fig. 3.6. The influence of suspended load and colloids on the REE concentrations of river water. (Elderfield et al. 1990)

Not just the pH but also the presence of colloidal phases in suspension can control the concentration of the rare earths in river water (Goldstein et al.1984; Sholkovitz and Elderfield 1988; Elderfield et al. 1990). River waters that are enriched in colloidal phases display higher REE concentrations than rivers that are poor in such phases. The influence of the suspended load (including colloidal phases) on REE concentration is shown in Fig. 3.6. This diagram shows the behaviour of the REE in river water in successive experiments. River water, which was filtered through <2µm and <0.45µm filters, shows concentrations that are almost identical to those of the original, untreated sample. Use of a <0.2µm filter reduces the REE concentration of the river water by about 60 to 80%. This means, therefore, that a significant portion of the REEs exist in suspended form in the <0.45µm fraction. REE concentration is seen to decrease with pore size. It is scarcely possible to filter off all such colloidal particles. Therefore, REE concentrations for river water always represent a product of mixing between dissolved ions and REE integrated into particles. Iron, in the form of iron hydroxide ($Fe(OH)_3$) and its other oxides and hydroxides, is the single most important component of colloidal particles. The concentration of Fe in river water is also controlled by pH. The amount of colloidal particles and hence the Fe content is directly correlatable with the concentration of the REE (Fig. 3.7).

River water, which has been filtered through <2µm filters, shows an almost identical REE pattern to that of the unfiltered sample, while increasing fractionation of the REE can be observed with further decreases in pore size. Depletion of light and heavy REE can be seen in the fine fraction.   These HREE

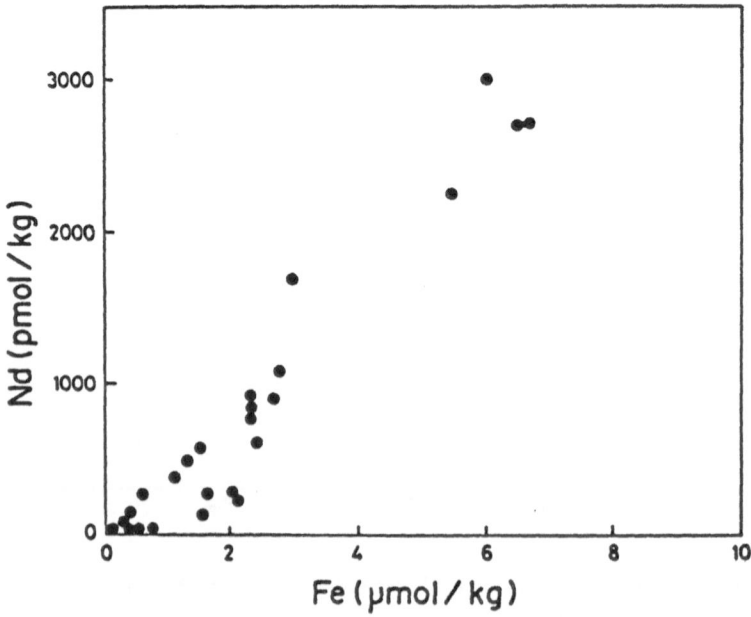

Fig. 3.7. Relationship between Fe and Nd concentrations in filtered river water. (Elderfield et al. 1990)

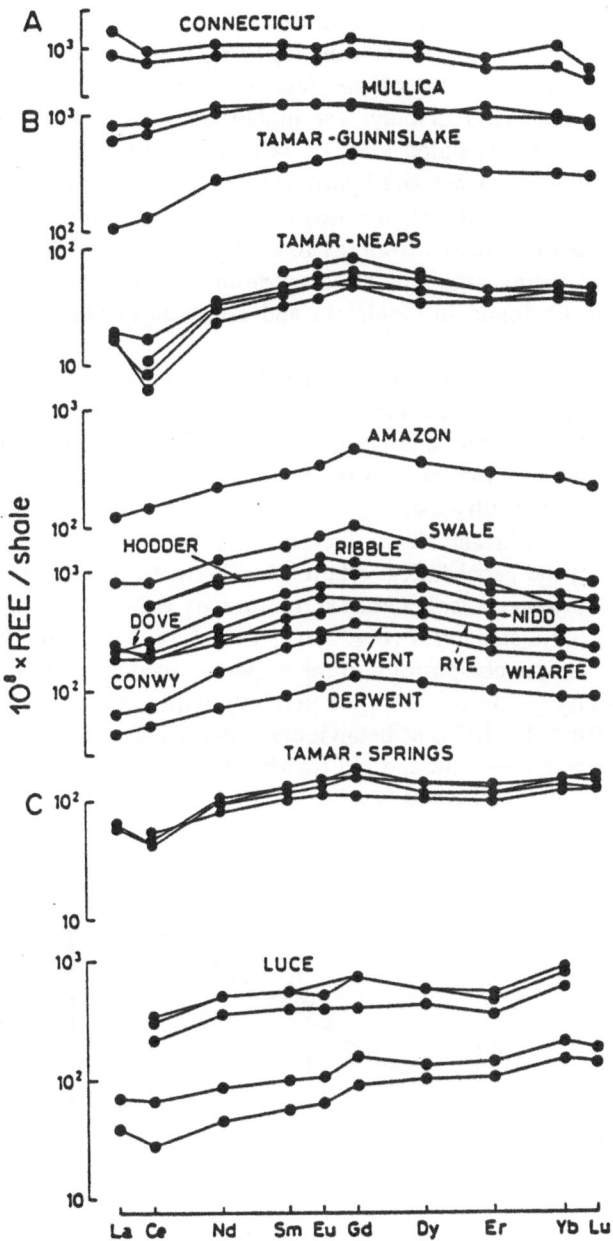

Fig. 3.8. REE distributions in river waters. (Elderfield et al. 1990)

and LREE appear to be preferentially enriched in the coarser particles. This fractionation mechanism has important consequences for the Nd isotope system and Sm/Nd ratio. High Sm/Nd ratios are to be expected in the dissolved fraction, conversely low Sm/Nd ratios are to be expected for the suspension. Goldstein and Jacobsen (1987) were able to demonstrate that $(Sm/Nd)^{water} / (Sm/Nd)^{susp.} > 1$ is the case for most of the river systems of the World. This ratio is also pH dependent. River water with pH>7 displays far higher Sm/Nd ratios in the dissolved fraction than in the suspended load. It is assumed that higher pH increases the stablility of the carbonate and hydroxide complexes of the heavy REE relative to the light REE. $^{147}Sm/^{144}Nd$ ratios determined on filtered river water display a large degree of variation, from 0.1 to 0.18. Most ratios lie around 0.12 which corresponds to the average crustal composition. Typical patterns of variation of the REE can be found in Goldstein and Jacobsen (1988a,b) and Elderfield et al. (1990; Fig. 3.8).

The investigations of Goldstein and Jacobsen (1987, 1988c) showed that the $^{143}Nd/^{144}Nd$ ratios of the suspended load and filtered water vary only little (Fig. 3.9). However, it was always possible to establish, that even filtered water could contain colloidal phases and so differences in the isotopic composition between dissolved and colloidal Nd might still exist.

The isotopic compositions of Sr and Nd in river water are inversely correlated with each other and reflect the geochemical relationship between Sr and Nd in crustal rocks. Rivers which drain ancient cratons show strongly radiogenic Sr and weakly radiogenic Nd isotopic signatures (Fig. 3.10). Therefore, the Sr and Nd isotope ratios reflect the isotopic composition of the region drained by the rivers. The isotope ratios of Sr and Nd do not correlate perfectly with each other. The poor correlation results from the different behaviours of the Rb-Sr and Sm-Nd isotopic systems during weathering. Sm and Nd fractionate with respect to each other relatively little compared with Rb and Sr.

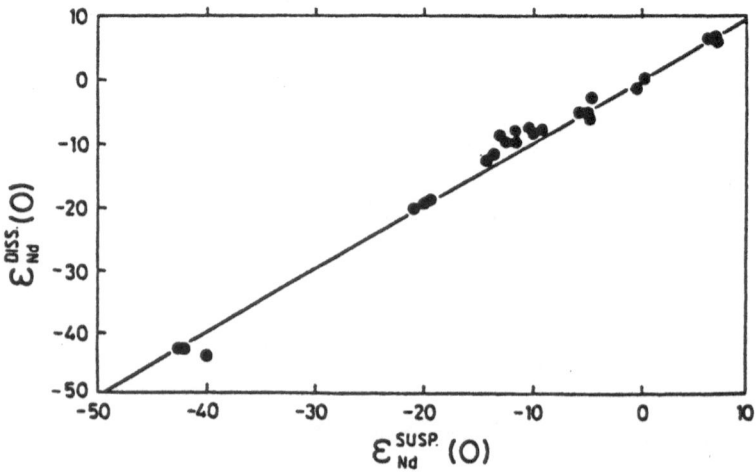

Fig. 3.9. Relationship between Nd isotopic composition of dissolved and suspended load. (Goldstein and Jacobsen 1987)

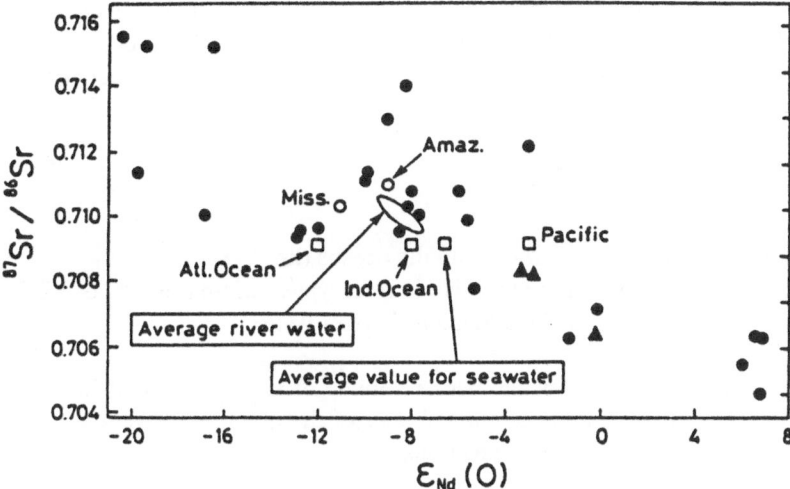

Fig. 3.10. Relationship between Sr and Nd isotopic composition of filtered river water. (Goldstein and Jacobsen 1987)

Rb and Sr show greater fractionation not only during weathering but even during magmatic processes in the crust (granite formation, melting of sediments, etc.). As well as this, the Rb/Sr ratios in the crust vary much more than Sm/Nd. This has the consequence that the $^{87}Sr/^{86}Sr$ ratios in the crust are more strongly variable than $^{143}Nd/^{144}Nd$ ratios.

Isotopic and elemental analyses from the most important river waters of the World make it possible to calculate mass balances for the ocean (for compilation see e.g. Goldstein and Jacobsen (1987) and Table 3.1):

example:
Calculation of the amount of Nd which is brought into the oceans annually by river transport and the average Nd isotopic composition of the load.

known:
-average Nd concentration in river water: 41 ppt $= 41*10^{-12}$ g/cm^3
-volume of water flowing annually into the World's oceans: 42,439 km^3/year

This allows us to calculate the amount of dissolved Nd that is brought into the oceans by rivers per year: $1.7*10^9$ g. River water, therefore, has an important influence on the concentration and isotopic composition of marine neodymium.

This calculated amount of dissolved Nd that is carried into the oceans by rivers is a maximum figure, which is not likely to correspond to the real value because dissolved rare earth elements in river water (including Sm and Nd) are unstable and liable to precipitate out of solution in various ways (formation of complexes with organic compounds, compounding with Fe-hydroxides, adsorption onto particles in suspension). Goldstein and Jacobsen (1987) have suggested the following equation to estimate the amount of dissolved Nd that is carried into the oceans by rivers:

$$J^{RW(eff.)} = \sum_i \emptyset_i M_i C(Nd)_i \qquad (I)$$

whereby: M_i=flow of river water into the oceans (km^3/year)
$C(Nd)_i$=concentration of Nd in river water before removal of REE
$\emptyset_i$=amount of Nd that remains in river water

The isotopic composition of the whole volume of river water that finds its way into the oceans is dependent on the volumes of the individual rivers and their Nd concentrations and isotopic compositions. Therefore:

$$\varepsilon = \frac{(X\varepsilon 1 + Y\varepsilon 2 + Z\varepsilon 3)}{(X + Y + Z)} \qquad (II)$$

where εi represents the isotopic composition and X, Y and Z the portion of the whole (X+Y+Z) flow. If equation (I) is to be used, the isotopic composition can be written as:

$$\varepsilon^{RW(eff.)} = \frac{\sum \emptyset_i M_i C(Nd)_i \varepsilon_i (Nd)}{J^{RW(eff.)}} \qquad (III)$$

where ε_i corresponds to the Nd isotopic composition in each river. The following equations assume that $\emptyset$ und $C(Nd)$ are relatively constant in all rivers. In this case, according to Goldstein and Jacobsen (1987), equation (I) can be modified as follows:

$$J^{RW(eff)} = \emptyset C(Nd)_i \sum_i M_i \qquad (IV)$$

the isotopic composition is:

$$\varepsilon^{RW(eff)} = \frac{\sum_i M_i \varepsilon_i}{\sum_i M_i} \qquad (V)$$

With the help of equation (V) an εNd value of -8.4 can be calculated for river waters (see Fig. 3.10). These sorts of estimates are important in order to calculate mass balances and the evolution of marine isotopic systems.

As has been common in such studies, the work of Goldstein and Jacobsen considered only the influence of dissolved elements in river water when estimating the marine element budget. In actual fact, some studies have shown that the marine budget may also be influenced by the suspended load transported into the oceans by the rivers. We have already discussed that particular matter carried by the rivers is significantly more enriched in Sm and Nd than filtered river water.

Table 3.1. Concentrations of the REE in rivers and seawater. (data from Goldstein and Jacobsen 1988)

sample	La	Ce	Nd	Sm	Eu	Dy	Er	Yb
susp. load [1]	39.6	80.9	36.4	6.9	1.43	4.18	1.98	1.6
diss. load [2]	30.8	64.5	40.9	10.8	2.66	11.5	8.46	6.1
eff. diss. load [2]	8.3	25.8	16.4	4.97	1.04	5.41	5.16	3.5
seawater [2]	5.3	1.8	3.5	0.65	0.16	1.2	1.2	1.2

[1] in ppm; [2] in ppt;

Leaching experiments show that large amounts of REE and Sr are to be found adsorbed onto particle surfaces or integrated in HCl soluble mineral phases (Nägler et al. 1993). Before leaching, the particles have a $^{87}Sr/^{86}Sr$ ratio of 0.72025 and a Sr concentration of 135 ppm (SL1, Fig. 3.11). After leaching in 1N strength HCl, the leachate has a Sr isotopic composition of 0.71041 and a Sr concentration of 410 ppm (L1). This experiment clearly demonstrates that a large portion of the Sr is situated in exchangeable sites and/or associated with a soluble phase. The Sr isotopic composition of the leachate is similar to that for the filtered river water, which gives 0.71026. The dissolved Sr could possibly represent adsorbed river water, a mixture of this with ancient marine carbonate Sr, or other Sr-rich components such as phosphates. The particle residue (R) must, according to mass balance, lie on the straight line, having a more radiogenic signature and a lower Sr concentration than the untreated particle fraction.

The exchange behavior of the elements adsorbed onto the particles with elements from the surrounding fluid is made clearer by the following experiment: Particles in suspension were dried in a beaker, then mixed with seawater and following that subjected to the same already described experimental procedures. The non-leached particles now showed a $^{87}Sr/^{86}Sr$ ratio of 0.71825 and Sr concentration of 154 ppm (SL2). Obviously, exchange with seawater, which had a $^{87}Sr/^{86}Sr$ ratio of 0.709034, must have served to decrease the Sr isotopic composition of these particles. The isotopic composition of the leachate (L2) was that of the seawater. Similar exchange behaviour can be observed with Nd. The REE are seven times more concentrated at exchangeable sites than within the particle itself. It is to be expected that these sorts of exchange processes on particle surfaces must lead to fractionation in seawater.

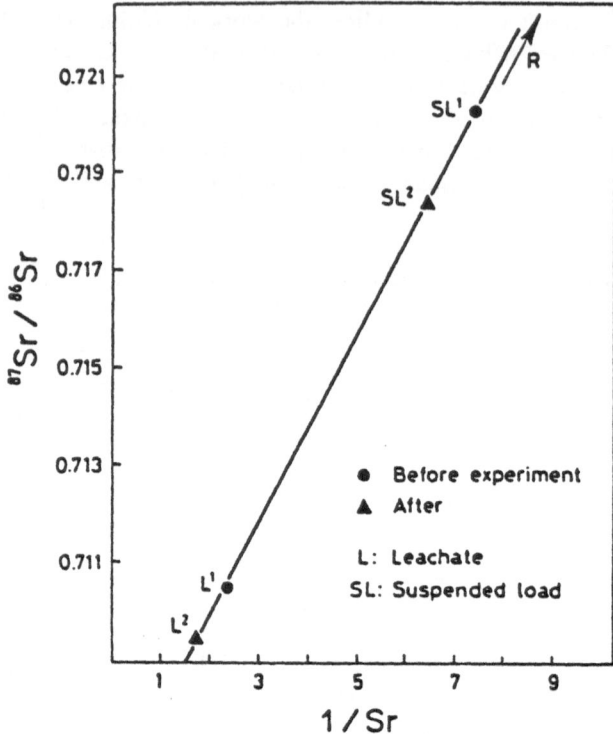

Fig. 3.11. Sr mixing diagram illustrating the exchange behavior of Sr adsorbed on particulates of suspended load (SL1) with filtered river and seawater. L1: 1N HCl leachate of SL1. SL2: exchange of dried suspended load (SL1) with seawater. L2: 1N HCl leachate of SL2, R: residue. (Nägler et al. 1993)

Suspended loads show depletion in the heavy REE relative to average shale, which is commonly used to represent average upper crust, and large variation in the light REE, which fluctuate around average shale concentrations. The amount of REE dissolved in river waters is far lower. They show a slight enrichment in the heavy REE (Fig. 3.8). The portion provided by river water to the budget of Sr and Nd to the oceans must be significantly greater than previously calculated. This requires of course that some of the suspended load discharging into the ocean goes into solution and does not sink to the seafloor as particulate matter (see also Sect. 5.6).

3.3 References

Elderfield H, Upstill-Goddard R, Sholkovitz ER (1990), The rare earth elements in rivers, estuaries, and coastal seas and their significance to the composition of ocean waters. Geochim Cosmochim Acta, 54: 971-991

Goldstein, SJ, Jacobsen SB (1987) The Nd and Sr isotopic systematics of River-Water dissolved material: implications for the sources of Nd and Sr in seawater. Chem Geol (Isotope Geoscience Section), 66: 245-272

Goldstein SJ, Jacobsen SB (1988a) Rare earth elements in river waters. Earth Planet Sci Lett, 89: 35-47

Goldstein SJ, Jacobsen SB (1988b) REE in the Great Whale River estuary, northwest Quebec. Earth Planet Sci Lett, 88: 241-252

Goldstein SJ, Jacobsen SB (1988c) Nd and Sr isotopic systematics of river water suspended material: implications for crustal evolution. Earth Planet Sci Lett, 87: 249-265

Goldstein SL, O'Nions RK, Hamilton PJ (1984) A Sm-Nd isotopic study of atmospheric dusts and particulates from major river systems. Earth Planet Sci Lett, 70: 221-236

Nägler T, Stille P, Chaudhuri S, Clauer N (1993) A Sr -Nd study on dissolved and suspended loads of Mississippi River waters with implications on global mass balance calculations. VII Meet Eur Union Geosci, April 4-8, Strasbourg, Terra Abstr, p344

Sholkovitz ER, Elderfield H (1988) The cycling of dissolved rare earth elements in Chesapeake Bay. Global Geochem Cycles 2: 157-176

Sigg L, Stumm W (1989) Aquatische Chemie. Verlag der Fachvereine, Zürich

Tricca A (1997) Transport mechanisms of trace elements in surface and ground water: Sr, Nd, U and rare earth elements evidence. PhD Thesis, Univ. Strasbourg

4 Isotope Geochemistry in the Environment

The introduction of anthropogenic, geogenic and pedogenic, organic and inorganic contaminants or poisons into our biosphere has been greatly accelerated due to recent human activity. Substances, which do not appear in nature, are being increasingly introduced into the natural cycles that link the continents, oceans and atmosphere together. This increased influx of abiotic material and anthropogenically induced, accelerated turn over of natural poisons places a burden on the environment, in particular the surface environment.

In order to characterize and understand the various cycles every source, transport route and transformation process must be determined for each and every substance. This subject is discussed in greater detail in Kummert and Stumm (1989) and Sigg and Stumm (1989). Surface water (rain, spring, pore and river waters) plays a significant role in the regulation of these material cycles. A schematic diagram (Fig. 4.1) gives us an idea of how material ends up in flowing, surface water.

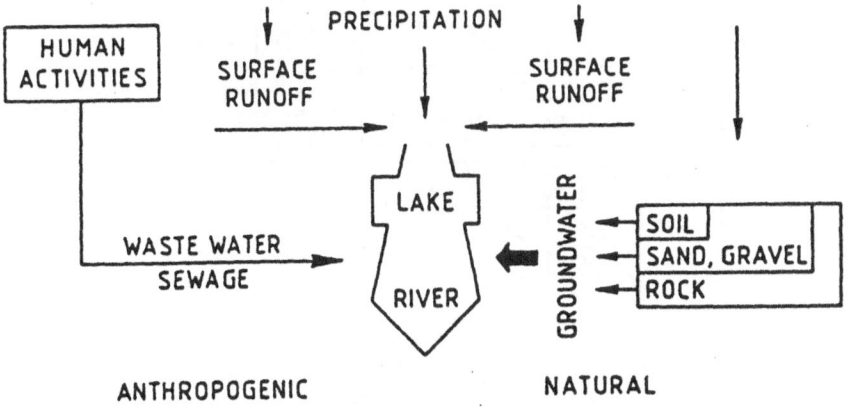

Fig. 4.1. Transport routes of anthropogenic and pedogenic contaminants in the flowing, surface water system. (Zobrist 1983)

Depending on the size and nature of the populous, waste water may make up between 2 and 20% of the flowing water. More than half the precipitation that falls on land, i.e. rain or melt water, filters away through the underlying soil through sand, gravel and rock before making it back to the surface by way of springs.

Trickling of contaminants through rocks and soil may lead to an increase in the concentration of certain substances, which are the products of chemical weathering. These sorts of weathering products make their appearance if soil or rock forming minerals react with acid rain or with naturally occurring carbonic acid from within the humus layer. After strong rains or rapid melting of snow, a large amount of water may remain on the surface. In this way, depending on the type of surface or the population density, the amount and influence of anthropogenic substances may increase substantially.

In a pioneering study, Murozumi et al. (1969) analyzed various samples of Greenland snow specifically for their lead concentrations. They reached the conclusion that Pb concentrations in Arctic precipitation had increased from pre-industrial time values of 1 pg Pb/g to 200 pg Pb/g at the present day. Despite some criticism of their interpretation at the time and the inability of subsequent workers to avoid problems of analytic precision and procedural blank Pb contamination, their data remain unchallenged and have been further supported since (e.g. see Wolff and Peel 1985). It is probable that the very first Pb contamination in the air derives from Roman times. Now, 2000 years later, it seems likely that "not even 2% of Pb in the global troposphere is from natural sources" (Boutron and Patterson 1986). This can be estimated by comparing Pb concentrations in Arctic

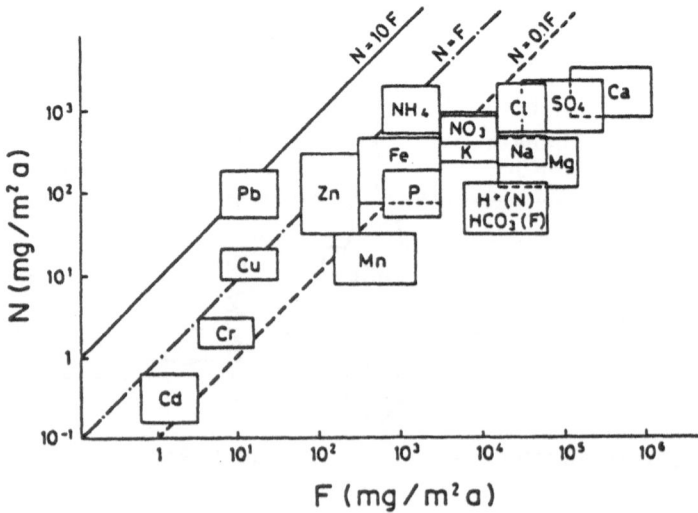

Fig. 4.2. Comparison of air borne fluxes (N:precipitation) with contaminant river fluxes (F; Zobrist 1983)

snow with concentrations of silicate dust and sulfur, which are considered to approximate fluxes of soil dust and volcanic fall out respectively, which are considered to be the main sources of natural lead in precipitation.

Zobrist (1983) investigated the fluxes of airborne material in detail. He was able to show that not only lead but also several other heavy metals including cadmium, copper and zinc are found in greater concentrations in rain water than would be expected, considering natural cycles (Fig. 4.2). The input of lead into the oceans through precipitation is actually greater than the input by way of rivers. The relative fluxes for ammonium, zinc and copper are about the same for both fluvial and atmospheric sources. For other substances, atmospheric input is outweighed by that from flowing surface water.

The transport of undesirable substances in dissolved form or by way of a colloidal phase is usually dependent upon physico-chemical conditions, e.g. pH. In the next part, we will discuss the origin, transport and exchange behavior of heavy metals and other trace elements, from the perspective of isotope geochemistry, as they make their way into the rivers and groundwaters from both anthropogenic and natural, geogenic sources, and during migration through the soil. Although isotope geochemistry has great potential for solving environmentally orientated problems, this approach has only been applied in a few specific cases. Isotope geochemistry allows us to characterize element cycling in the environment, to identify point sources of contamination and to observe exchange processes between fluid and solid phases. These applications will undoubtedly become clearer with carefully chosen examples.

4.1 Heavy Metals in the Environment

Lead is important when we consider the migration behavior of heavy metals. There are two reasons for this:

1. Our environment is heavily burdened by lead. Since the beginning of industrialization, the concentration of Pb in the atmosphere of the northern hemisphere has increased by a factor of one hundred. Today, almost all atmospheric lead comes from industrial sources. By far the largest contribution comes from the exhausts of motor vehicles as a byproduct of the combustion of petrol where lead is used as an anti-knocking agent. As a result of rain, snow or mist, lead may be rapidly taken up into the atmosphere, making its way into the soil and surface water and eventually into the biologic cycle.

2. Lead has several isotopes. As industrial lead is likely to have various characteristic Pb isotopic compositions, depending on how it was produced and released, industrial contamination can be identified. Mass spectrometry allows us

not only to differentiate between industrial and natural lead but can help trace contamination also. For this purpose, $^{206}Pb/^{207}Pb$, $^{207}Pb/^{204}Pb$ and $^{206}Pb/^{204}Pb$ ratios can be used.

Industrial lead comes from ore bodies, whose Pb isotopic composition differs markedly from that of average crustal rock. During the formation of lead ore bodies, uranium is separated from lead due to its starkly contrasting geochemical behavior. This has the consequence that ^{206}Pb and ^{207}Pb, which are the decay products of ^{238}U and ^{235}U respectively, are no longer produced after the formation of the lead ore body (U/Pb=0). Most ore bodies that exist today have therefore lower $^{206}Pb/^{207}Pb$, $^{206}Pb/^{204}Pb$ and $^{207}Pb/^{204}Pb$ isotopic ratios than average crustal · rocks in which lead can still be found together with uranium and where both ^{206}Pb and ^{207}Pb are still being produced.

Fig. 4.3 gives an overview of Pb isotopic compositions of important Pb ore bodies that are being used for industrial purposes today. The Broken Hill lead ores from New South Wales, Australia are found hosted by greywackes and are 1600-1700 million years old. They contain Pb with low $^{206}Pb/^{207}Pb$ ratios of around 1.039. The lead ores of Mount Isa, Queensland are also of the same age and give a similar $^{206}Pb/^{207}Pb$ ratio of 1.042. Precambrian lead sulphide, 1000 to 1300 million years old, from the Grenville Province in Canada yields higher $^{206}Pb/^{207}Pb$ ratios, which vary between 1.077 and 1.104.

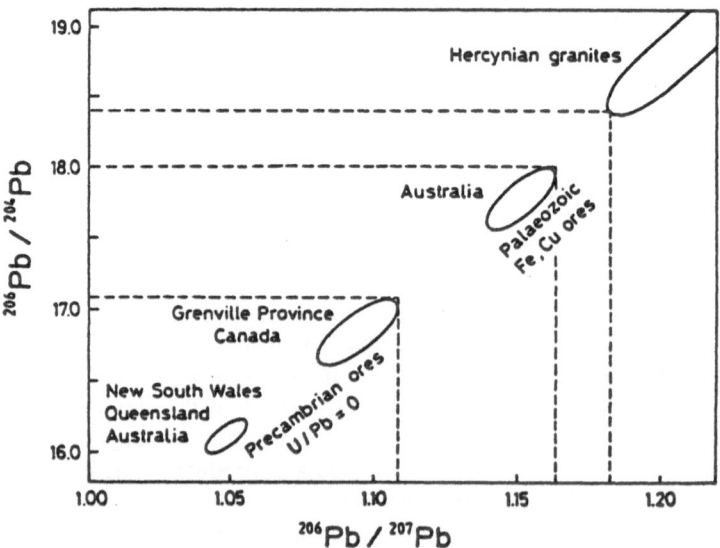

Fig. 4.3. Pb isotopic compositions of important Pb ore bodies that are being used for industrial purposes today. For comparison the natural lead originating from Hercynian granitic rocks have higher $^{206}Pb/^{207}Pb$ and $^{206}Pb/^{204}Pb$ ratios. (modified after Elbaz-Poulichet et al. 1986)

Paleozoic sulfides from Australia display even higher ratios of 1.14 to 1.16. All these Precambrian lead sulphide ores, whose Pb isotopic compositions have been more or less frozen by the removal of uranium (U/Pb=0) have been used in the manufacture of petrol. Their characteristic Pb isotopic signatures can now be identified worldwide.

At the end of the sixties some people started to think about the application of the Pb isotopic system to the solution of environment related problems. At that time it was not at all clear how much the Pb isotopic composition varied in the various environmental systems such as soil, water, air, plants as well as in petrol and in other industrial products and to what extent these differences could be distinguished using a mass spectrometer. The work of Ault et al. (1970) was pioneering in this respect. They were among the first to investigate Pb isotopic compositions in the environment. Although the mass spectrometer precision at that time can scarcely be compared with precision today, it is worthwhile taking a closer look at some of the results of their investigations.

Ault et al. were able to demonstrate that the $^{206}Pb/^{204}Pb$ ratios of particles in the air over various cities in America could vary significantly both within the city and between cities. This made it possible to identify point sources of lead for the first time. An interesting case is that of the important traffic route between Philadelphia and New York. It was here that the first attempt was made to determine the influence of vehicle derived lead particles on soils and vegetation. Various soils, grasses and leaves were analyzed in the study. Their Pb isotopic compositions gave a homogeneous distribution pattern. Low $^{206}Pb/^{204}Pb$ ratios were found near the freeway, which then rose with increasing distance from the freeway (Fig. 4.4). The lead from petrol is therefore less radiogenic than the 'natural' background lead in this region. The Pb anomalies show that some of the larger lead particles derived from cars can be transported up to 800 to 1000 m.

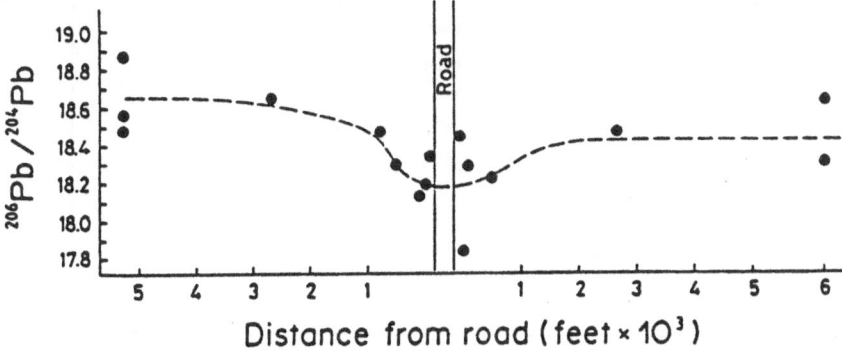

Fig. 4.4. Lead isotopic composition of soils shown against distance from the freeway. (Ault et al. 1970)

The Pb isotopic analyses imply that only the uppermost centimeters of the soil are strongly enriched in industrial lead. Although the work from Ault et al. showed the way, not only to source detection for contaminating heavy metals but also to the reconstruction of heavy metal circulation in the biosphere, it was not until the eighties that the Pb isotope method came to be used in environmental research. The strengths of the Pb isotope systems in demonstrating atmospheric Pb contamination and differentiating various sources of contamination are clear in the work of Sturges and Barrie (1987), in which it was shown that the industrial lead used in Canada and in the USA comes from very different sources.

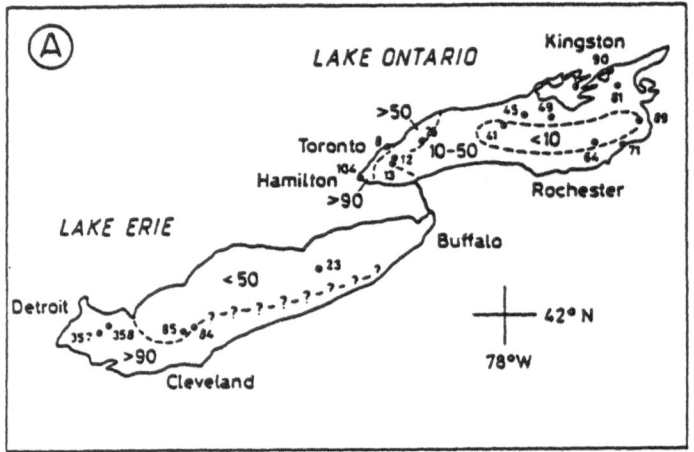

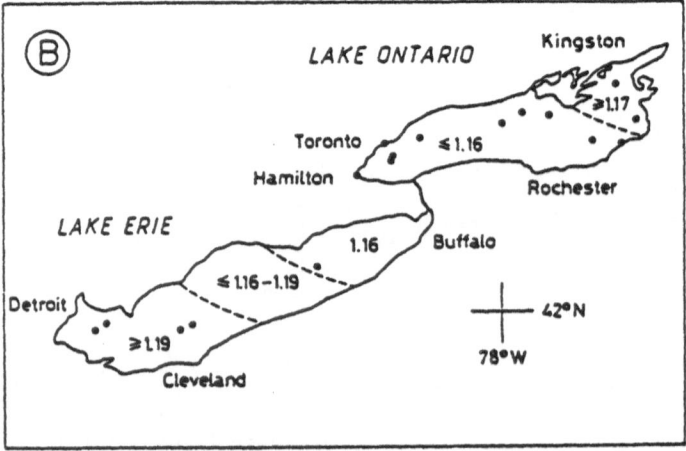

Fig. 4.5. (A) Distribution of Pb concentrations (pmol kg^{-1}) and (B) ^{206}Pb/^{207}Pb ratios in surface waters from the great lakes (Lake Ontario, Lake Erie). (Flegal et al. 1989)

In this way, aerosols from Canada and the USA could be differentiated giving $^{206}Pb/^{207}Pb$ ratios of 1.151 ± 0.010 and 1.221 ± 0.009, respectively. Such fine differences were not discernable at the time of Ault et al. in 1970. This study provided the impetus for further Pb isotope studies, concentrating on the identification of industrial lead in soils and lakes, which are derived directly from the atmosphere.

Flegal et al. (1989) investigated the Pb isotopic compositions of surface water from the great lakes (Lake Ontario, Lake Erie; Fig. 4.5). The lead concentrations of dissolved lead in surface waters are highly variable and enriched in Pb close to industrial centres. The $^{206}Pb/^{207}Pb$ ratios are equally variable and reflect the various source regions which come into play. Most of the $^{206}Pb/^{207}Pb$ ratios of the central and western basins of Lake Ontario lie between 1.15 and 1.16 and are therefore consistent with ratios for the industrial centre of Toronto (1.15+/-0.01). At two localities at the eastern end of Lake Ontario and in the port of Hamilton, anomalously high $^{206}Pb/^{207}Pb$ ratios were measured of >1.18. These values lie between values for Canadian and American aerosols and point to a mixing of the two. Lake Erie allows us to recognize a trend in isotopic compositions from West to East. The highest ratios (1.19) are found in the West and the lowest (1.16) in the East. This means that the amount of industrial American lead decreases from West to East. Regression calculations indicate that the lead in these waters comes from two sources only. A natural geogenic source of lead cannot be identified. Virtually all the lead that is dissolved in the surface waters of Lake Erie and Ontario is therefore of industrial origin and originates from both Canada and the USA.

Various studies have shown that the residence time of lead dissolved in the surface waters of a lake is only on the order of 4 to 7 days (Sigg, 1985). The dissolved lead is quickly adsorbed onto particles or is complexed with organic molecules. This complexation and adsorption strongly control the budget of lead in surface waters. The short residence time of lead has the consequence that Pb isotopic composition at any one time may give an idea of lake circulation and climatic conditions (wind direction in particular).

The transport of industrial lead by river systems is discussed in detail by Elbaz-Poulichet et al. (1984, 1986). The difference between petrol lead and natural lead is quite straight forward in France as 99% of lead used is imported, mostly from Canadian and Australian Precambrian ore bodies (Fig. 4.3). In some parts of France, lead from other sources is used, mainly for the manufacture and treatment of iron, steel and copper. This lead can be recognized by its relatively high $^{206}Pb/^{207}Pb$ ratios. The natural lead of France comes from 300 million year old Hercynian granite plutons, which have $^{206}Pb/^{207}Pb$ ratios of between 1.18 and 1.20 (Fig. 4.3). More recent sediments originate, to a high degree, from these parent rocks and possess therefore a similar Pb isotopic composition. In this study, the authors attempted to define the various points of origin for the lead which can be found dissolved or in suspension in the most important French river systems

(Seine, Loire, Garonne-Gironde, Rhone) and also to ascertain the amount of contamination due to industrial lead.

In order to categorize the various lead components, industrial and pre-industrial, according to their isotopic compositions, Pb isotopic ratios of aerosols and some of the older delta sediments of French rivers were analyzed. The isotopic compositions are displayed in Fig. 4.6. Pre-industrial delta sediments show the highest $^{206}Pb/^{207}Pb$ ratios and lie in the region of Paleozoic granites and sediments (Fig. 4.3). The aerosols which contain an industrial Pb component are the least radiogenic and show similar Pb isotopic compositions to Precambrian ore bodies. Isotopic analyses of a recent soil profile, which covers the period from 1800 to 1978, demonstrates beyond doubt the increasing influence of industrial lead in the environment. The $^{206}Pb/^{207}Pb$ ratios of all the largest French rivers lie between those of aerosols and pre-industrial sediments. The lowest, most industrially influenced isotopic ratios are shown by the Seine and the Loire, followed by the Garonne and the Rhone. The Pb ratios for the suspended load in the delta regions are lower and are therefore the most seriously contaminated.

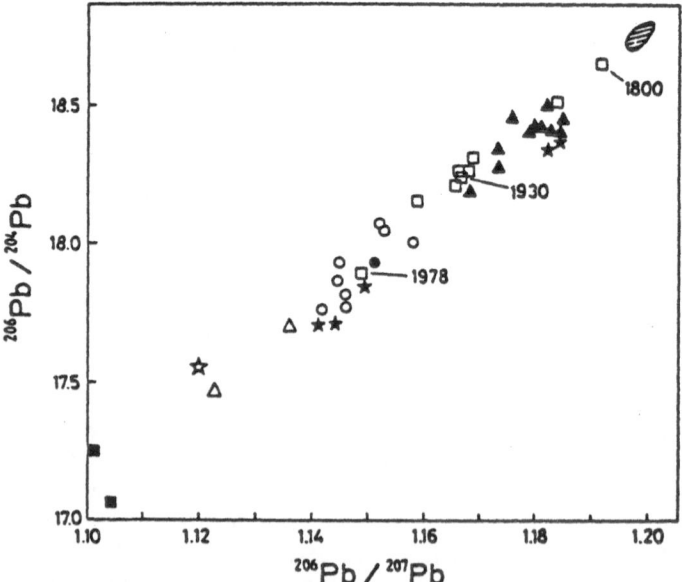

Fig. 4.6. Pb isotopic compositions of aerosols, suspended loads and older delta sediments of French rivers and lake sediments. Squares: drilled lake sediments from the Pavin crater lake (Massif Central) covering the period between 1800 and 1978. Hatched field: pre-industrial river sediments. Filled triangles: Garonne and Gironde. Filled stars: Rhone. Open circles and triangles: Loire. Filled circle: Seine. Filled squares: aerosols. Open star: filter dust (Paris). (Elbaz-Poulichet et al. 1996)

If one assumes a ratio of 1.1 for anthropogenic Pb and a ratio of 1.2 for natural Pb, it is possible to estimate, within a fair margin of error, the relative amount of anthopogenic Pb which is transported into the rivers. This would be 45-80% for the Seine, 40-70% for the Loire, 25-43% for the Garonne and 15-28% for the Rhone. Thus, it can be shown that the Seine is the most contaminated river of the four. Annually, the Seine transports 75-140 tons of anthropogenic Pb as suspended load into the Atlantic. The Rhone transports 70-125 tons, the Garonne 40-75 tons and the Loire 35-60. These enormous amounts of Pb, which are carried each year by these rivers into the oceans are directly related to the main industrial agglomerations of France. For example, 30% of the French population live close to the Seine river basin, along with 40% of the whole commercial activity.

The transport of industrial Pb in a soil down to the groundwater table is discussed in Erel et al. (1990). This study looks at the application of the Pb isotopic method to the investigation of transport mechanisms and the reconstruction of the migration routes of heavy metals from the atmosphere into groundwater. The area of investigation is located in a secluded valley in the Yosemite national park (USA). Snow and river water samples were collected as well as samples from a 30 cm deep soil profile, which reached down to the groundwater table.

The soil samples were all treated with 1N HNO_3 in order to leach the most easily soluble phases containing lead, iron, manganese and organic material from the soil. Erel et al. name this component "labile reservoir" (LR). This leaching experiment is not only effective in mobilizing cations that are adsorbed on the surfaces of particles but also in dissolving iron hydroxides and manganese oxides. The lead, iron and manganese fractions that remain in the soil after this treatment are located mainly in the magmatic minerals of weathered granite and represent the insoluble residue (NLR). The original $^{206}Pb/^{207}Pb$ ratios of fresh, unweathered

Table 4.1. Pb concentrations and $^{206}Pb/^{207}Pb$ ratios in soluble and insoluble reservoirs of the soil (after Erel et al. 1990)

| depth (cm) | soluble (LR) | | insoluble (NLR) | |
	Pb (ppm)	$^{206}Pb/^{207}Pb$	Pb(ppm)	$^{206}Pb/^{207}Pb$
0	157	1.190	–	–
0 - 0.5	81	1.192	44	1.210
0.5 - 1.0	37	1.217	--	–
2.0 - 2.5	15	1.300	--	--
4.5 - 5.2	14	1.280	72	1.211
14.5 - 15.0	13	1.295	82	1.215
35 - 45	8	--	--	–
fresh granite			22	1.212

granite are around 1.212. The Pb isotopic compositions of the insoluble residue give values varying between 1.21 and 1.215 and are therefore identical to the unweathered granite (Table 4.1). The soluble labile component (LR) shows not only a trend in Pb isotope composition but also in Pb concentration with increasing depth in the soil profile. The Pb concentration decreases with increasing depth while the $^{206}Pb/^{207}Pb$ ratios increase. On the basis of isotope geochemistry, two distinct regions can be recognized in the soil profile:

In the uppermost centimeters of the soil profile $^{206}Pb/^{207}Pb$ ratios between 1.19 and 1.217 can be determined for the LR. These values are close to those of the atmosphere in this region (average $^{206}Pb/^{207}Pb$ since 1870: 1.17; present-day: 1.22). It can therefore be assumed that the uppermost centimeters of the soil profile contain a significant component of industrial lead. Deeper in the profile, the ratios vary around 1.3, which is far higher than that characteristic for industrial lead and is also higher than that of fresh, unweathered granite or that of the insoluble residue (NLR; 1.22). Erel et al. consider that this part is not influenced by industrial lead but by the weathering of accessory minerals such as apatite, allanite and monazite with high U/Pb ratios and therefore high Pb isotopic compositions. This free, radiogenic lead was released into the labile reservoir (LR) of the rock system.

The soil profile was also measured for its organic carbon content (Table 4.2). The concentrations of lead and organic carbon allow us to recognize an anthropogenic as well as a geogenic trend (Fig. 4.7).

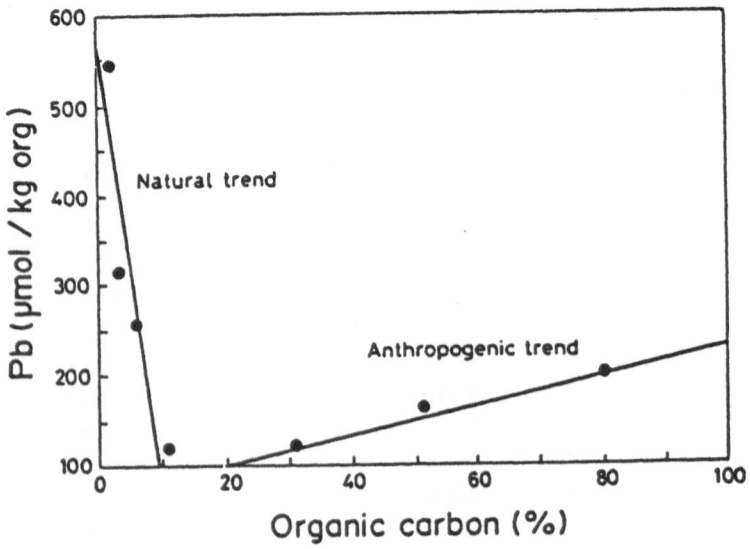

Fig. 4.7. Relationship between Pb concentration and organic matter in soils. (Erel et al. 1990)

Increasing depth sees not only a decrease in organic carbon but also of lead. At 2 cm, the Pb concentrations reach a minimum, then Pb content begins to increase along with organic carbon. This variation can be viewed as the result of natural weathering processes without any recognizable contribution from industrial lead. This natural, geogenic trend relates to the strong binding behaviour associated with lead and organic carbon, which only ceases where the amount of organic carbon is reduced due to oxidation. Oxidation of organic carbon releases acids, which attack accessory U-rich minerals and allow for the mobilization of radiogenic lead in the soil.

Table 4.2. Percentage proportions of water, organic matter and soluble components in the soil (after Erel et al. 1990)

depth(cm)	water (%)	soluble (%)	Org. (%)
0	40	90	80
0-0.5	68	54.3	50.7
0.5-1.0	60	31.0	31.0
2.0-2.5	37	11.3	6.0
4.5-5.2	14	18.7	11.0
14.5-15.0	27	6.8	2.3
35-45	14	8.3	2.6

Two important aspects have come to light here. First, the close relationship between Pb concentration and organic carbon, (Corg) and second, the strong enrichment of industrial Pb in the uppermost parts of the soil profile. If we know the isotopic ratio of atmospheric Pb for the last 130 years ($^{206}Pb/^{207}Pb$: 1.17), the concentration of industrial Pb in snow melt that seeps into the ground (Table 4.3), and we can determine the isotopic compositions of natural Pb in the soil profile and in unweathered granite ($^{206}Pb/^{207}Pb$: 1.21), we can then go on to calculate the relative contribution of industrial Pb to the uppermost two centimeters of the soil.

In order to observe the exchange of industrial Pb down to the groundwater table, Erel et al. investigated not only snow and river water (SW) but also groundwater (GW) and porewater (SM; Fig. 5.8) for their Pb isotopic compositions. Snow and river water show identical isotopic compositions that also correspond with those of the atmosphere in this region. The Pb concentrations are much higher in snow than in river water. This allows us to assume that the larger part of the lead (90%) does not go into the rivers after snow melting but instead seeps directly into the soil.

Although the chemical composition of river water is very similar to that of groundwater it is clear that they have different sources of lead on the basis of their different Pb isotopic compositions (Fig. 4.8). Groundwater has the same isotopic composition as that found in pore water (SM). This isotopic composition shows no similarity with the insoluble residue (NLR), which is constant over the whole soil profile nor to that of the soluble component (LR), which shows an increase at the same depths as the groundwater. The intermediate isotopic composition of both ground and pore waters indicates that a certain amount of industrial lead has found its way into the groundwater. Assuming that there are two mixing end members, 1.2 for the industrial component and 1.29 for the soluble component (LR) close to that of both groundwater and pore water, allows us to calculate that about 15% of the lead comes from industrial sources.

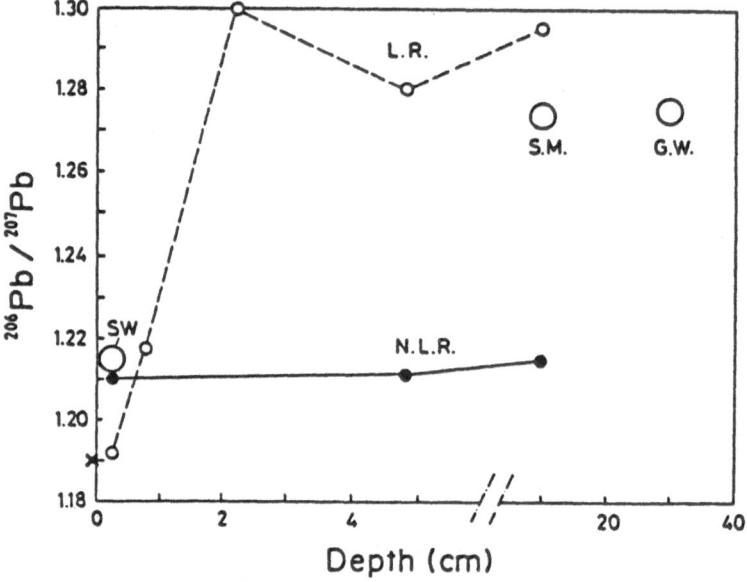

Fig. 4.8. $^{206}Pb/^{207}Pb$ of the soluble (LR) and insoluble (NLR) reservoirs as a function of depth in the soil profile. GW: groundwater; SW: river water; SM: interstitial water. (Erel et al. 1990)

Table 4.3. Pb concentrations (nmol/kg) and $^{206}Pb/^{207}Pb$ ratios in snow, river and groundwater. (after Erel et al. 1990)

water source	year	Pb (nmol kg-1)	($^{206}Pb/^{207}Pb$)
snow	1973	2.95	1.20
snow	1983	0.483	1.22
river	1972	0.068 - 0.082	----
river	1986	0.04	1.215
soil moisture	1972	33.82	1.274
groundwater	1986	0.072	1.275

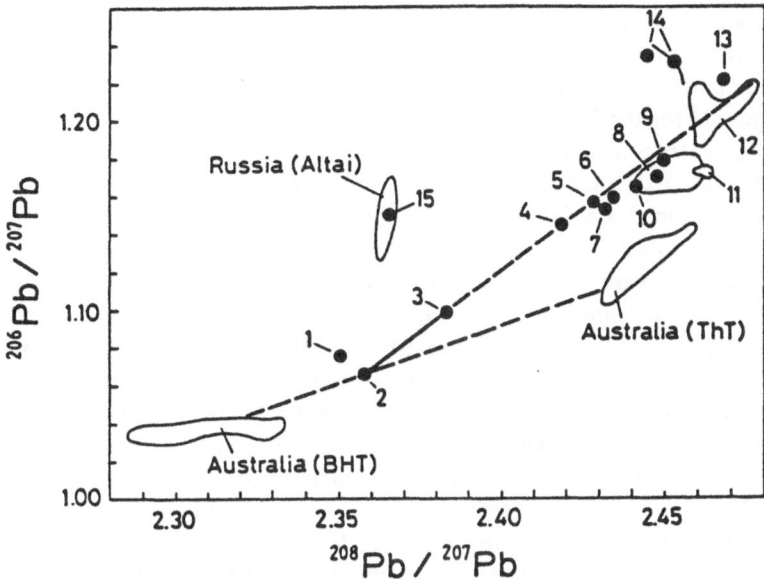

Fig. 4.9. Anthropogenic lead contamination in Munich (Germany) shown using $^{206}Pb/^{207}Pb$ and $^{208}Pb/^{207}Pb$ ratios. 1: sewage works; 2: petrol lead (1985-87); 3: air filter; 4: beer; 5: blood; 6: tap water; 7-9: older sewage works; 10: soil; 11: triassic Pb-Zn deposits; 12: carbon coal; 13: sediments of Isar river in Munich; 14: limestones of the Munich area; 15: sewage work deposits from 1976. (Horn et al. 1987)

An additional isotope study that may be of interest here was able to demonstrate anthropogenic lead contamination and the origin of this lead in a large European city (Munich, Germany; Horn et al. 1987). In this study, not only were Pb isotope analyses carried out on water, sediment and soil samples, but also on petrol, blood, beer and other materials such as air filter dust and coal. It is striking in the $^{206}Pb/^{207}Pb$ vs. $^{208}Pb/^{207}Pb$ diagram (Fig. 4.9) that almost all components go towards defining a mixing line, with carbon-coal on one side and petrol lead on the other defining the end members. The coal shows enhanced $^{206}Pb/^{207}Pb$ and $^{208}Pb/^{207}Pb$ isotopic ratios, while petrol lead used between 1985 and 1987 shows low lead isotope compositions. Lead in petrol which is an anti-knocking agent (lead tetramethyl) derives from two Australian localities, Broken-Hill and Thakaringa in the proportion 60:40.

The isotopic composition of the lead from the air filter can be looked at as a product of the mixing of lead from coal and petrol lead. This ought to be at least close to the truth as the air filter was located near a coal-heated power station. The isotopic ratio of blood lead lies likewise on the same mixing line: petrol lead - air filter - coal lead. From the isotope data it is easy to estimate that about 35% of the whole blood lead content comes from petrol. Beer lead shows somewhat lower lead isotopic ratios but lies close to blood lead. The proportion of petrol lead in beer lies around 40%. This relatively high amount is not surprising when we consider that the lead concentrations in cereal crops and beer are very low, around 0.04 bis 0.08 ppm. Tap water also lies on the mixing line but below that value which is characteristic for natural lead from the area around Munich and in the drainage area for the Munich water. This allows us to surmise that a part of the lead in tap water comes from anthropogenic sources. As the lead concentrations in tap water are extremely low, around 0.002 ppm, even the tiniest traces of anthropogenic lead are likely to have an influence and cause a shift in the lead isotopic composition.

Lead in the sewage from a sewage works in Munich (#1) shows a completely different isotopic composition than tap water (#6). Astoundingly however, Pb concentrations were found to be similarly low (0.0028 ppm). The Pb appears to have come predominantly from Australian sources. Pb isotopic compositions of older sewage works deposits (#7, 8 and 9) allow for the conclusion that this older 'sewage' Pb was derived from more local sources. Sewage deposits (#15) from the year 1976 show a Pb isotopic signature which can not relate to any lead ore deposit. Not even by mixing various ore bodies can such Pb isotope compositions be reached. The only known ore body on Earth which has a similar isotopic composition can be found in the Altay mountains (Russia). Was lead illegally imported into Germany and used in the Munich area up to 1976?

As we have seen, heavy metals are so much more enriched in an anthropogenically influenced environment than in nature. This makes it possible to use their concentrations and isotopic ratios in tracing both the source and extent of anthropogenic contamination. Lead is the prime example but is not always the only or the best possibility as seen in a recent multi-isotope (Pb, Sr, Nd) and REE study of a heavy metal contaminated soil (Steinmann and Stille 1997).

·One of the most sensitive indicators of the influence of sewage particles on the marine environment is silver (Ag) content. Silver concentrations are characteristically enhanced by up to 200 fold in sewage sludge relative to pristine marine sediments. Why sludge should be so enriched in silver must have something to do with its specialist uses in photography and electronics, on top of the fact that it is such as a rare element in nature: background Ag concentrations in marine sediments are only on the order of tens of ppb.

In some marine and aquatic environments, the concentrations of rare metals may be naturally enhanced as in the Black Sea, for example. Therefore, for a more definitive measure of contamination, we need to look at the isotopic ratios of rare metals as their high masses make kinetic isotopic fractionation of negligible importance. Anomalous isotopic ratios are most likely to occur due to the mixing of two end members, as with Pb, one anthropogenic and the other natural. The established use of Ag as a tracer for contamination enabled, first Esser and Turekian (1993), and then Ravizza and Bothner (1996) to test the potential of an even rarer element Osmium (Os) and its isotopic ratio $^{187}Os/^{186}Os$ for the same purpose. ^{187}Re decays to ^{187}Os as a result of beta decay with a half-life of 45.6 Ga. More details about the systematics of this decay can be found in Sect. 6.4. The system is essentially analogous to the Rb-Sr isotopic system in that crustal rocks are enriched in the heavier, radiogenic isotope ^{187}Os and mantle rocks contain more of the lighter isotope ^{186}Os. Osmium is mined from ultramafic, magmatic complexes and so anthropogenically derived Os ought to contain more of the lighter Os isotope than normal crustal rocks. In these, the first studies of their kind, Ag and Os concentrations and Os isotopic ratios were measured in short sediment cores from a series of stations close to the mouth of Boston Harbor and in the Cape Cod Bay, Massuchussetts. Two samples of sewage sludge were also analyzed in order to define the chemical characteristics of the metal source.

Sludges from the two different sewage deposits and one treatment plant contain osmium contents up to 10-40 times higher than background (i.e. pristine, pre-anthropogenic influence, marine sediments). Their Os isotope ratios are also very low and virtually indistinguishable from ore rocks as would be expected from anthropogenic Os. Os isotopic signatures from pristine sediments, sewage influenced sediments and sewage sludge can be compared in Table 4.4.

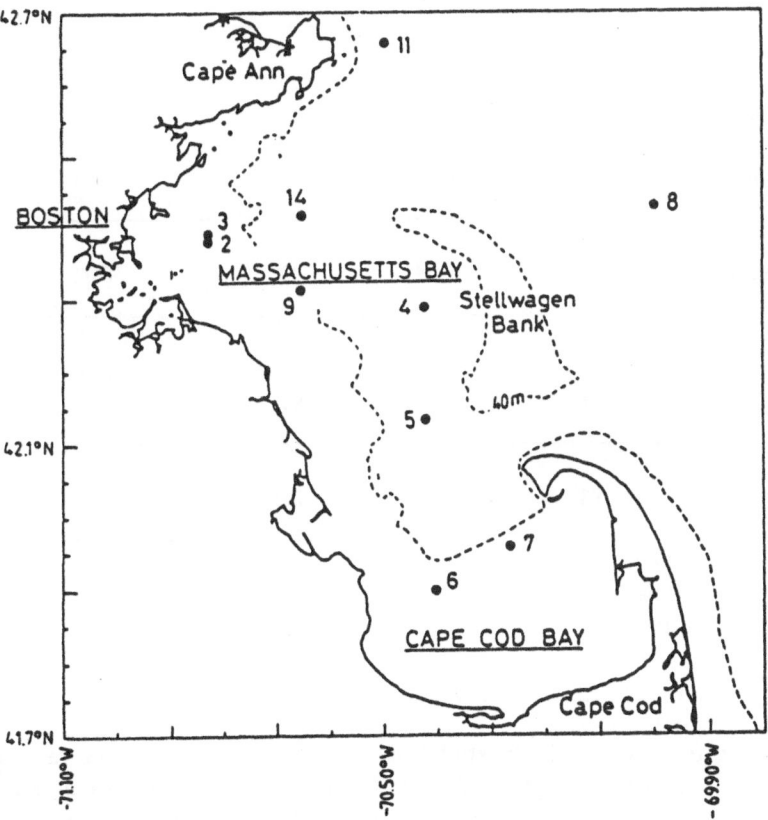

Fig. 4.10. Sampling locations for Os isotopic analysis as a pollutant tracer. (Ravizza and Bothner 1996)

Table 4.4. Os and Ag data from Boston Harbor, Massachusetts Bay, Cape Cod Bay sediments and Boston sewage sludge compared with the Bushveld ores, South Africa. Surface sediment from sampling stations close to the outlet into Boston Harbor are clearly contaminated by anthropogenically produced osmium (Ravizza and Bothner 1996).

Sample	C organic %	Ag (μg/g)	Os (pg/g)	$^{187}Os / ^{186}Os$
Station 8				
0-0.5 cm	1.7	0.14	94	8.42
26-28 cm	1.5	0.06	88	9.11
56-58cm	1.5	0.07	108	8.10
Station 7				
0-0.5 cm	2.6	0.49	81	7.39
3.0-3.5 cm	2.4	0.42	64	8.65
34-36 cm	1.9	0.07	58	8.68
Station 6				
1.0-1.5 cm	1.9	0.56	56	7.97
3.0-3.5 cm	2.4	0.89	84	7.52
6.0-7.0 cm	1.4	0.36	40	8.06
52-54 cm	0.56	0.03	22	8.73
Station 5				
2.5-3.0 cm	2.7	0.39	79	8.62
Station 4				
5.0-6.0 cm	2.4	0.29	238	3.50
32-34 cm	1.8	0.12	91	6.71
Boston Harbor				
7.0-9.0 cm	5.1	6.7	286	2.95
Sewage sludge 1980			4010	1.42
Sewage sludge 1994			1310	1.27
Treatment plant 1972			570	1.53
Bushveld ores				1.413-1.509

One interesting aspect of this study was the switch in Os isotope ratio in the sludge between 1980 and 1994. The 1980 sludge shows a similar isotopic ratio, i.e. 1.42, to the Bushveld ultrabasic ore complex, South Africa, which is mined for osmium. However since 1980, the isotopic ratio has changed and now stands at 1.27, much lower than that of the Bushveld complex. The two natural osmium end members: continental crust and mantle stand at 10-11 and 1.1, respectively and so this lower ratio must derive from a more primitive source. The only candidate ore body seems to be the Siberian ores. It could be that Russia has become the dominant exporter of platinum group elements to the USA. As such information is difficult to obtain, and the major source of anthropogenic osmium is not known well enough, the authors do not choose to speculate further. However, the potential of the heavy metal approach to pollution research is clear.

4.2 Migration of Contaminated River Water into Groundwater Systems

The enrichment of various toxic substances in groundwater is a direct result of the infiltration of contaminated river water. Radon and oxygen isotope analyses carried out on river and groundwater have made it possible for rates of exchange between surface and groundwater to be calculated (Siegenthaler and Shotterer 1977; Stichler et al. 1986; Hoehn and Gunten 1989).

^{222}Rn is a daughter product of the decay series of ^{238}U (Fig. 4.11) with a half-life of only 3.8 days. The enrichment of ^{222}Rn in the upper atmosphere is mainly a consequence of diffusive emanation from U bearing minerals in granitic rocks, the rate of emanation being significantly dependent on the microstructure of the rock. In granular aquifers the intensity of radiation and rate of transport of radon is strongly influenced by the grain size distribution, weathering of grain surfaces and porewater content. Since the noble gas radon is almost chemically inert, it can migrate unhindered and diffuse from particle surfaces into the nannopores between grains. In water saturated aquifers, radon transport is strongly dependent on the speed of groundwater currents. Groundwater is particularly enriched, whereas surface waters are strongly depleted in radon. This concentration difference permits the determination of infiltration or exfiltration rates using radon as a natural tracer.

Age dating using radionuclides is based on one hand on the decay of the radionuclides (equations IV, V; Sect. 1.2), and on the other hand on the increasing concentration of the daughter product, which will after a certain amount of time exist in equilibrium with the mother nuclide (N), i.e. secular equilibrium. Equations (IV) and (V) may produce the following equation for the increase in amount of the daughter isotope (D):

$$D = N_0(1-e^{-\lambda t}) \qquad \text{(I)}$$

for radon:

$$A_t = A_e(1-e^{-\lambda t}) \qquad \text{(II)}$$

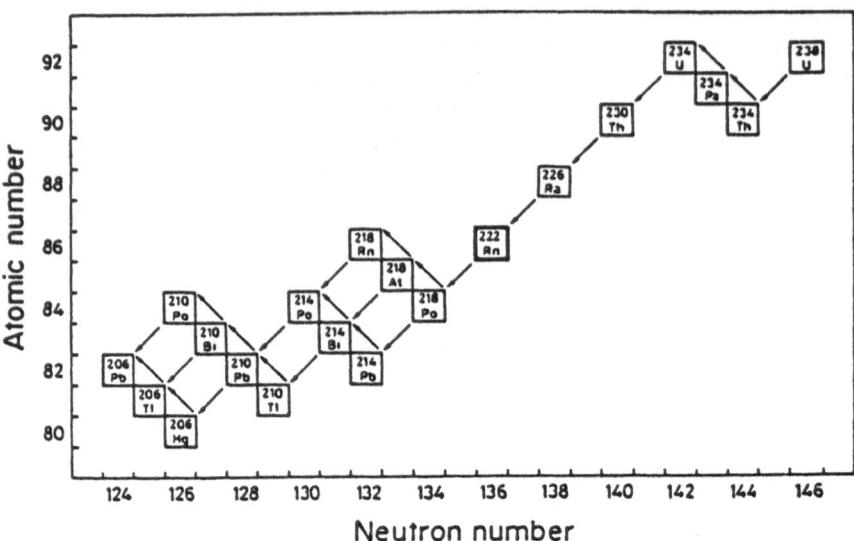

Fig. 4.11. The radioactive decay series of ^{238}U.

where A_t: activity of the daughter nuclide at time t. The activity is proportional to the concentration ($A_x = \beta.N_x$; N_x = number of atoms of the nuclide x). The activity is given in Becquerels (Bq; disintegrations per time unit). A_e: activity of the daughter nuclide in equilibrium with the mother nuclide ("secular equilibrium": activity of mother = activity of daughter). λ: decay constant (for ^{222}Rn = 0.18 day^{-1})

Equation II can be used for the calculation of the residence time:

$$t = \left(\frac{1}{\lambda}\right)\ln\left(\frac{A_e}{A_e - A_t}\right)$$

(II)

The concentration of radon (A_t) in the uppermost levels of the aquifer and the migration of Rn poor river waters (0.1-0.4 BqL^{-1}; Becquerel/litre) increases with increasing residence time until dynamic equilibrium is reached with the mother isotope ("steady state"). It takes about 20 days for dynamic equilibrium to be reached between radium (which is also a product of the ^{238}U decay series) and radon. After this amount of time, Rn has almost reached the concentration of its mother nuclide (Krishnaswami et al. 1982). Equation III is only valid if the infiltrating water is not already mixed with respect to Rn enriched groundwater. Hydrologic investigations show that very little exchange is likely to have taken place in the uppermost levels of the aquifer between river and groundwater.

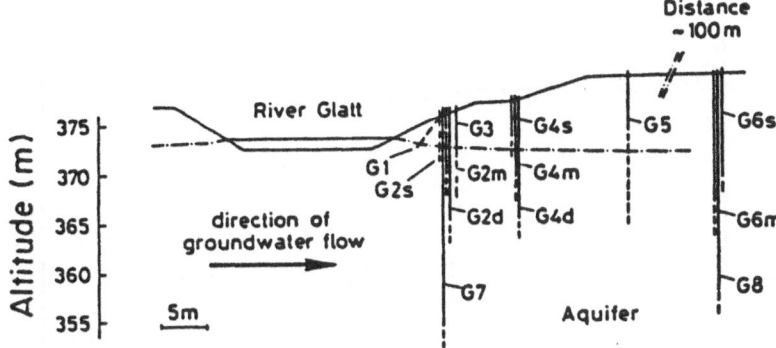

Fig. 4.12. Profile through the area of investigation shows the river Glatt (Switzerland), the direction of groundwater flow and the different wells G1-G8 used for groundwater sampling. (Hoehn and von Gunten 1989)

Table 4.5. Radon concentrations and residence times. (after Hoehn and von Gunten 1989)

well	distance from river (meters)	Radon concentrations $(Bq\,L^{-1})$	residence time (days)
G1	2.5	4.7 ±1.5	1.8 ± 0.06
G2s	5.0	6.7 ± 1.0	2.8 ± 0.04
G3	7.0	11 ± 3	6.4 ± 1.4
G4	14	9.1 ± 1.7	4.2 ± 0.08
G5	26	13 ± 2	7.9 ± 1.3
G6s	100	17 ± 1	>15

Fig. 5.12 shows the area of investigation: Glatt (in Switzerland; Hoehn and von Gunten 1989). Water samples were collected from the springs G1-G6. The radon concentrations determined from these waters are displayed in Table 5.5.

The concentrations increase with increasing distance from the river as might be expected. Assuming that the sample G6s, which was the furthest sample away from the Glatt river, has reached a concentration of radon characteristic of dynamic equilibrium (Ae=17 BqL⁻¹), makes it possible to calculate residence times for migrating waters in the uppermost layer of the aquifer of between 1.8 and more than 15 days. Fig. 4.13 shows a linear relationship between residence time and the distance between the spring and the river. This permits us to estimate that the speed of run through in this upper groundwater level is about 4.6 m/day.

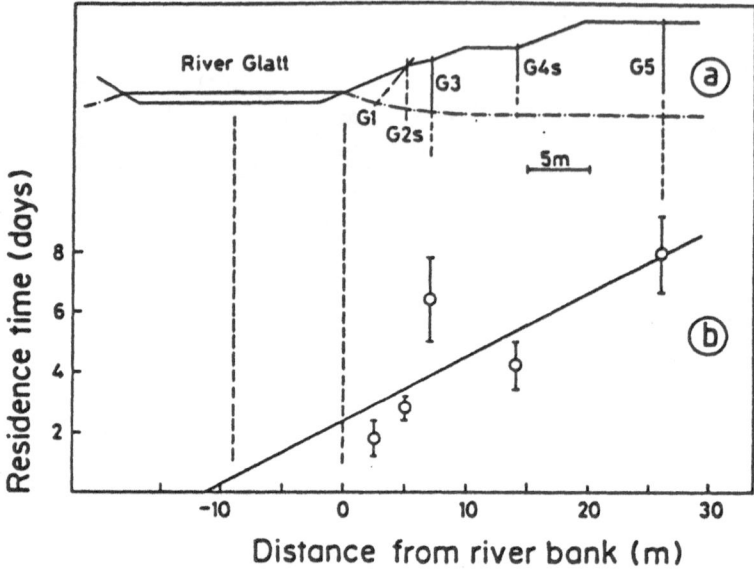

Fig. 4.13. Radon dating of groundwater. Groundwater residence time between Glatt river and observation wells G1-G5. (Hoehn and von Gunten 1989)

Sample G3 suggests a longer residence time, which may imply some exchange with water from deeper levels. The balance between river and groundwater exchange as well as the rate of exchange between the two systems can also be estimated with the help of the oxygen isotope system. The study of Stichler et al. (1986) carried out on the River Danube near Passau illustrates this well. Due to the lack of groundwater, river water, which makes its way into the uppermost layers of the groundwater table, is used as drinking water. Water chemistry must be constantly monitored and it is of the utmost importance that the rate of exchange between river and groundwater be known. Fig 4.14 gives the localities of investigation springs. Spring OW contains pure groundwater that has not experienced exchange with the river Danube. The other springs contain water that is affected by exchange with river water to various extents. Fig. 4.15 shows this exchange between river and groundwater schematically. Variations in oxygen isotopic composition are displayed for OW and the Danube between 1980-1982 in Fig. 4.16. From this we can detect significantly different isotopic compositions for the various water systems. Groundwater clearly shows higher $\delta^{18}O$ values than river water. Annual fluctuations can be recognized for Danube river water but these fluctuations do not manifest themselves in groundwater. The isotope characteristics of groundwater and Danube river water can be explained as follows: 1. Groundwater is more strongly enriched in ^{18}O than river water as it derives from pre-Alp precipitation, which is less strongly fractionated than the true Alpine precipitation which is the source of the river water.

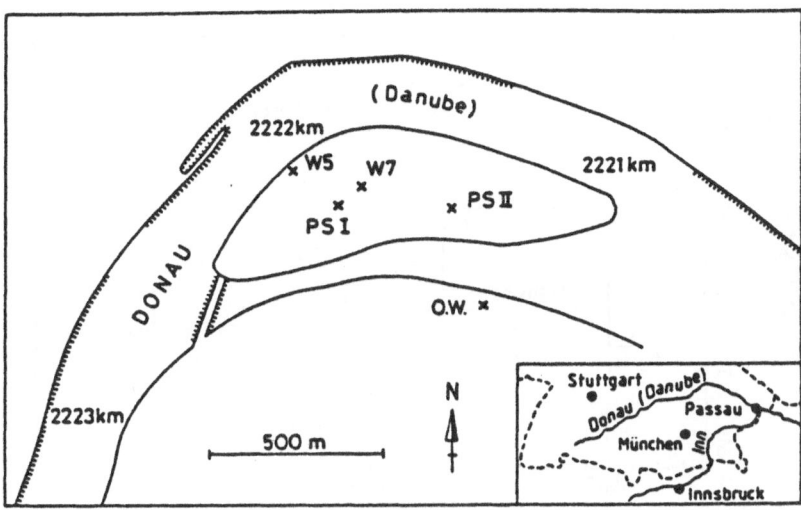

Fig. 4.14. Observation wells (PS, W5, W7, OW) close to the Danube river. (Stichler et al. 1986)

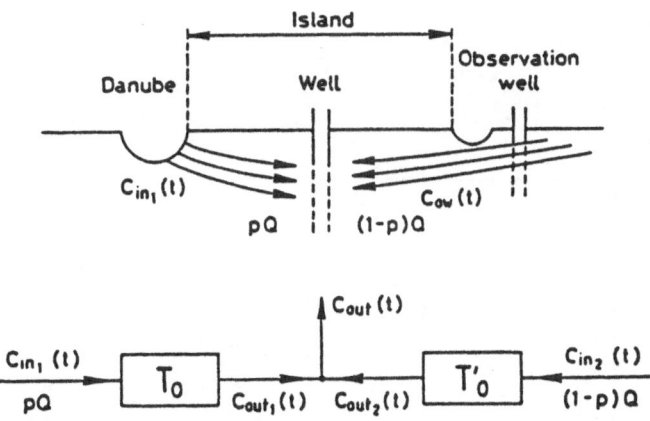

Fig. 4.15. Schematic flow model illustrating the hydrologic situation in the area of investigation. $C_{in}1(t)$: Tracer flux from Danube river; $C_{ow}(t)$: Tracer flux from groundwater observed in observation well OW; $C_{out}(t)$: Tracer concentration in observation wells PS, W5 or W7; p: fraction of the total flow rate Q; T=V/Q=average transit time of the groundwater (T_0 for Danube river; T_0' for groundwater; V:volume of the migrating groundwater. (Stichler et al. 1986)

In the Alpine region, both snow and rain waters are strongly fractionated, isotopically speaking, because of repeated evaporation and condensation events as altitude is gained (see Fig. 2.9). 2. The fluctuations in the isotopic composition of river water can be directly related to climatically forced temperature fluctuations. Enhanced Summer temperatures mean that ^{16}O is preferentially incorporated into

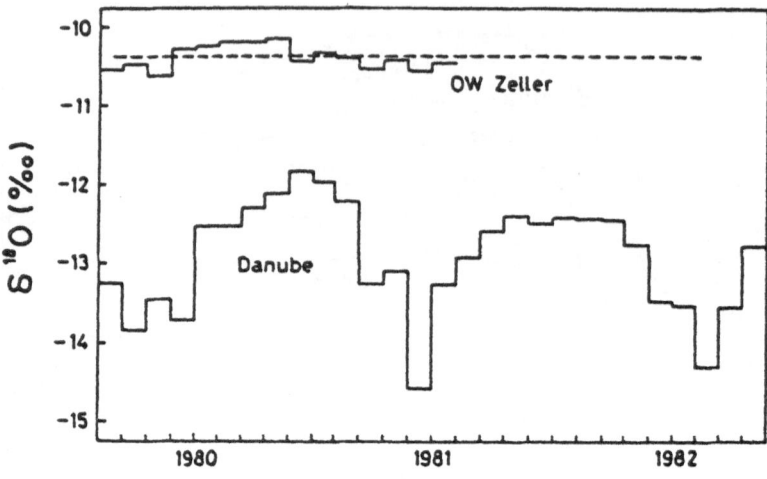

Fig. 4.16. Variation in $\delta^{18}O$ in groundwater and Danube river water from 1979-1982. (Stichler et al. 1986)

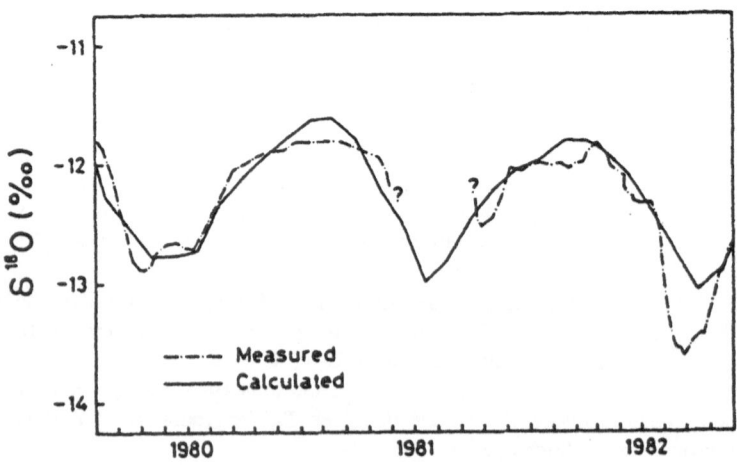

Fig. 4.17. Variation in $\delta^{18}O$ in the groundwater of one of the observation wells from 1979-1982. (Stichler et al. 1986)

the evaporated phase, while the heavier isotope ^{18}O remains in the remaining river water leading to higher $\delta^{18}O$ during the Summer (see also Sect. 2.4). The significant differences between the isotopic compositions of groundwater and river water allow us to quantify the sequence of exchange between the two systems. Let us look again at the schematic diagram of Fig. 4.15. The trace element concentrations in drinking water from a freshwater spring are dependent on the trace concentrations in both groundwater and river water, the relative amount that both water systems make up of the whole and the flow rate into the drinking water reservoir. The flow rate at such a spring is defined as follows: $Q=V/To$ (where V=volume of water and To is the transit time). The flow rate Q can be worked out from both the flow rates of the river water (p*Q) and groundwater (1-p*)Q added together (see Fig. 4.15). The various springs must therefore show an isotopic composition that should be the product of mixing between groundwater and river water. Fig. 4.17 displays the variation in isotopic composition of such a spring from 1980-1982. The $\delta^{18}O$ values are slightly higher than those for river water but do show the same time dependent fluctuations. This implies that river water makes up a large part of drinking water. Calculation of trace element concentrations in drinking water:

The concentration of an element X in a mixture can be described as follows: $X_m =$ p Xa + (1-p) Xb where Xa and Xb are the concentrations of element X in components A and B respectively and p is their relative amount in the mixture. This general equation can be used directly in the case of drinking water.

$$C_{out}(t) = p \, C_{out1}(t) + (1-p) \, C_{out \, 2}(t)$$

If we include data on oxygen isotopic composition we find that for p:

$$P = \frac{\delta^{18}O_x - \delta^{18}O_{ow}}{\delta^{18}O_d - \delta^{18}O_{ow}}$$

where:
$\delta^{18}O_d$= isotopic composition of Danube river water
$\delta^{18}O_{ow}$= isotopic composition of groundwater
$\delta^{18}O_x$= isotopic composition of the mixture

With the measured values for $\delta^{18}O$, we can calculate that the amount of river water in the drinking fountains W5, W7, PSI and PSII 77 can be up to 96% (Table 4.6). Mathematical flow models which simulate the isotopic variation in the drinking water allow us to make a number of predictions regarding the time taken for a toxic element to get into the drinking water system. Such models allow us similarly to represent the concentration of this element (C_{out}) in drinking water, and its concentration in riverwater (C_{out1}) as a function of time. The relationship of the concentration of an element with time is shown for four different springs in Fig. 4.18.

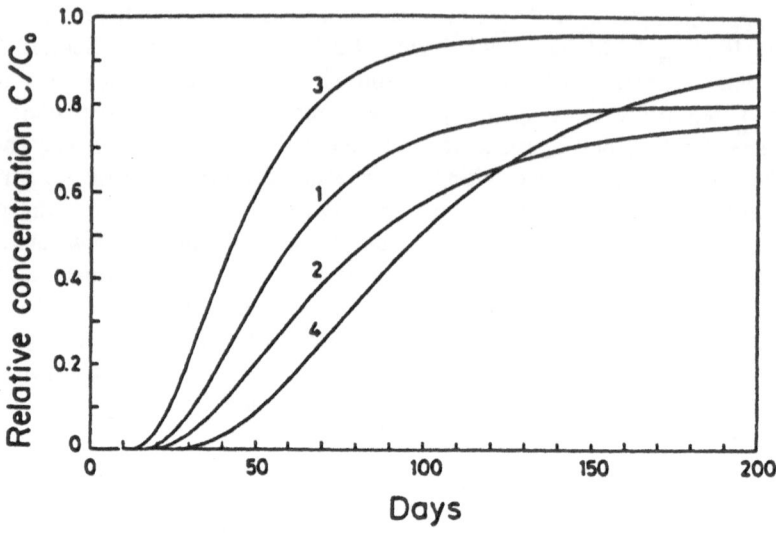

Fig. 4.18. Time dependent variation of tracer concentration. C_0: tracer concentration in the Danube river; C: tracer concentration in the different observation wells. (Stichler et al. 1986)

Table 4.6. Average $\delta^{18}O$ values (‰, rel. SMOW) and p factors. (after Stichler et al. 1986)

well	$\delta^{18}O(d)$	$\delta^{18}O(ow)$	$\delta^{18}O(x)$	p
W.5	-12.8	-10.4	-12.7	0.96
W.7	-12.8	-10.4	-12.6	0.92
PS.I	-12.9	-10.4	-12.4	0.80
PS.II	-13.0	-10.4	-12.4	0.77

Such models can also be applied whenever an accident threatens the quality of the water supply. However, they must be used with great care as the mathematic simulations relate to the laws of hydro-mechanics and do not concern themselves in the main with chemical exchange processes between water and organic matter, sand, mud or clay mineral particles.

4.3 Anthropogenic Contamination in River Water Observed Using Isotope Methods (R. Rhine)

More than 72 km^3 of the River Rhine flows into the North Sea annually making it the fortieth greatest river in the World. However, it is one of the most polluted. Its hydrochemical evolution through the years 1975-1984 was comprehensively discussed by Van der Veijden and Middleburg (1989) on the basis of major and trace element data. The evolution of the O, C and Sr isotopic systems from source to mouth were not investigated any more closely until Bühl et al. (1991) and Tricca (1993). These studies are described here. Fig. 4.19 shows the Rhine valley, various tributaries and the sample collection points of Bühl et al. (1991).

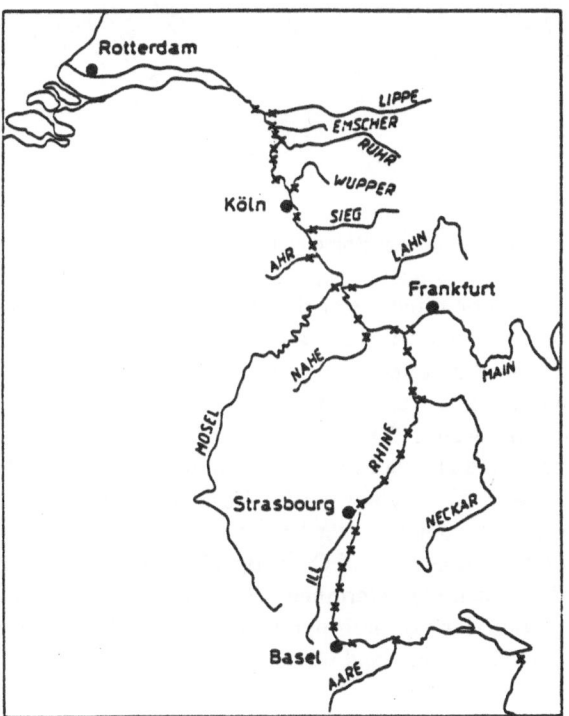

Fig. 4.19. Rhine valley with its various tributaries and the sample collection points. (Bühl et al. 1991)

4.3.1 Oxygen Isotopes

The Alpine Rhine displays typical $\delta^{18}O$ values for Alpine meltwater of around -12.5‰ SMOW (Fig. 4.20). This value has also been obtained from the Aare river

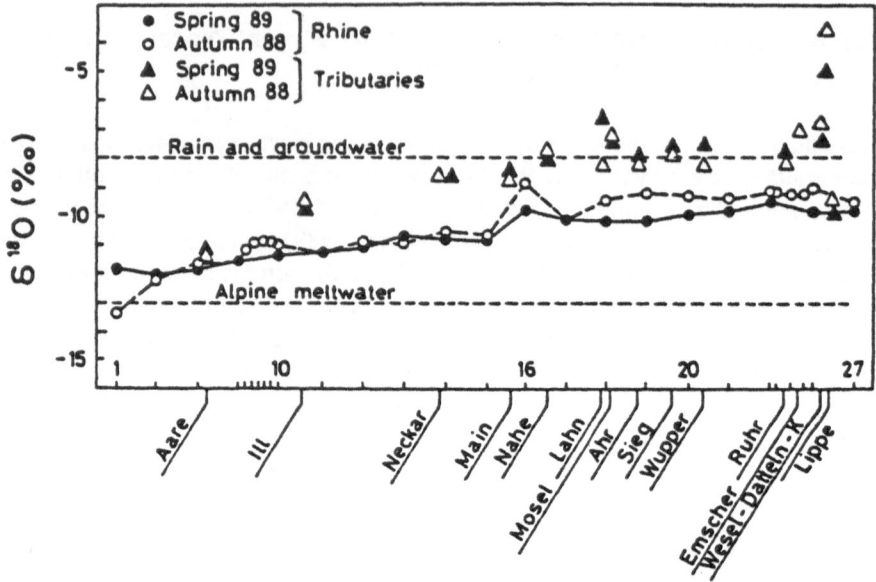

Fig. 4.20. $\delta^{18}O$ values in the river Rhine and in its tributaries. (Bühl et al. 1991)

(Switzerland), a major tributary of the Rhine. Rain and groundwater from pre-Alpine, or low lying regions, show a relative enrichment in $\delta^{18}O$ of about 8‰, something that is seen reflected in almost all tributaries of this region. Mass balance calculations allow us to establish that 30 to 50 % of Rhine water has its origin in Alpine meltwater. This is true even of Rhine water from the lower valley and has been confirmed by various hydrologic investigations. What stands out are the extremely 'heavy' $\delta^{18}O$ values for the Wesel-Datteln- Hamm canal. The reason for this is that heated, cooling water from a high temperature nuclear reactor (this leads to a strong fractionation of O isotopes) is piped into this canal. With the exception of this canal, the determined $\delta^{18}O$ values show a natural development down river without signs of anthropogenic influence as Alpine waters become progressively mixed with lowland water.

4.3.2 Strontium Isotopes

Sr isotopes as well as sodium and potassium contents clearly reflect anthropogenic influence. (Figs. 4.21, 4.22). A particularly strong rise in Na and K concentrations and $^{87}Sr/^{86}Sr$ ratios can be observed between localities 7 and 8, that is between Fessenheim and Breisach (Fig. 4.19). These increases show the influence of water from open cast mines or "pit water" and water used in the excavation process, "pit rinse water", from the Alsatian potassium mines flowing into the Rhine.

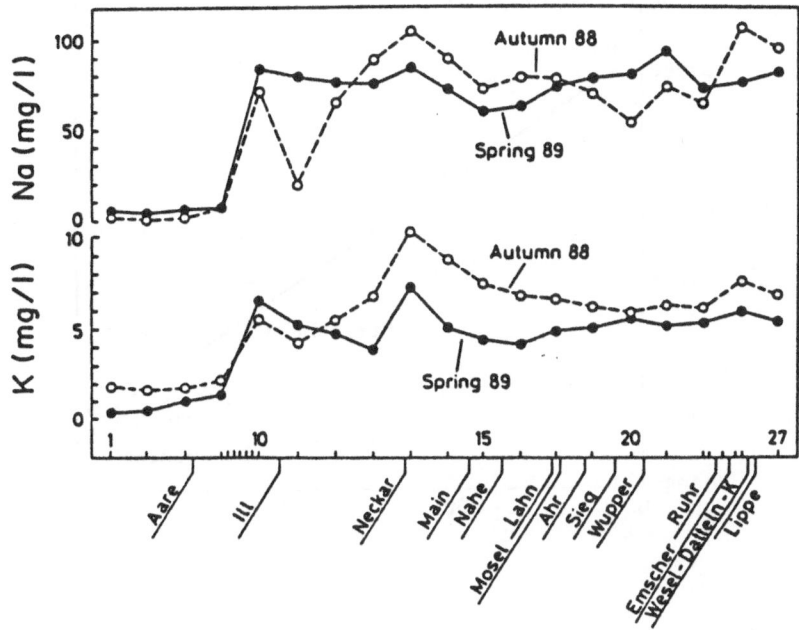

Fig. 4.21. Sodium and Potassium concentrations in the river Rhine. (Bühl et al. 1991)

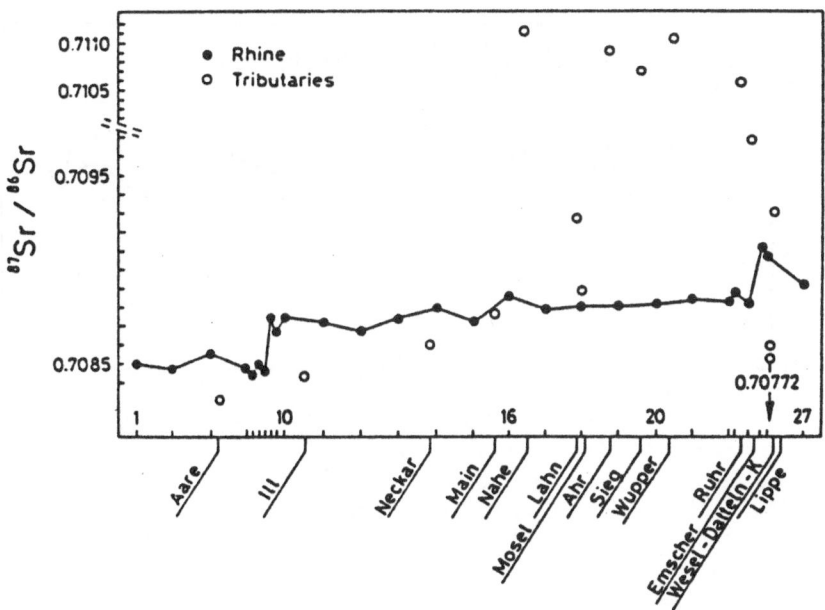

Fig. 4.22. Sr isotopic variation in the river Rhine and its tributaries. (Bühl et al. 1991)

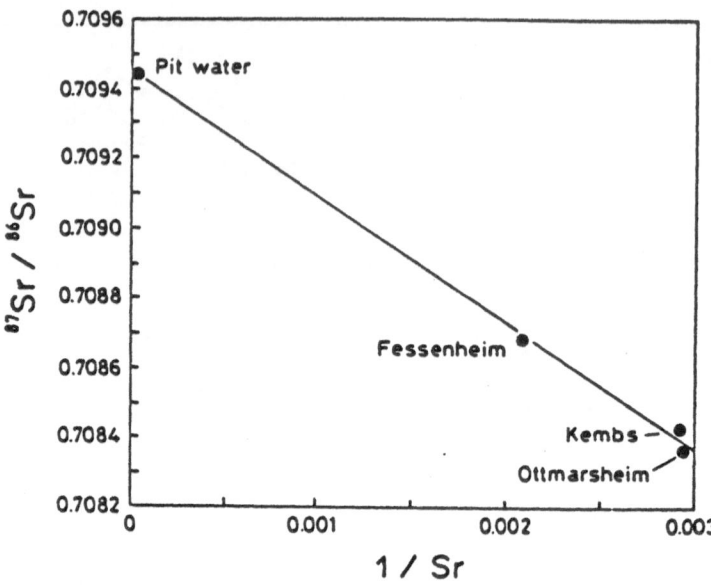

Fig. 4.23. The influence of pit rinse water from the potassium mines on Sr concentration and isotopic composition in the river Rhine. (Tricca 1997)

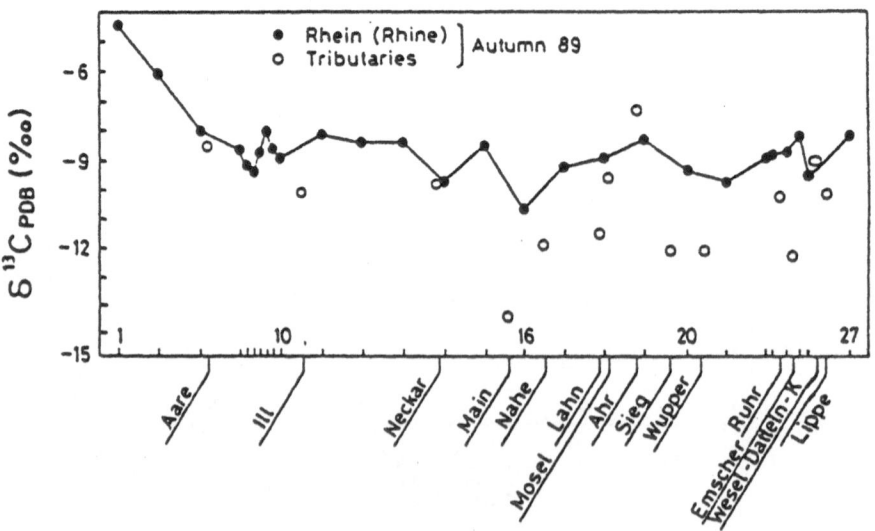

Fig. 4.24. The evolution of $\delta^{13}C$ in the river Rhine and its tributaries. (Bühl et al. 1991)

This addition of salts puts an extreme burden on the Rhine and has the consequence that both isotopic compositions and salt concentrations barely decrease on their way to the mouth of the Rhine. Additional input of salts can be observed at the neck of the Rhine in the industrial region of Mannheim-Ludwigshafen. The influence of pit rinse water from the potassium mines on the chemistry of the Rhine was discussed by Tricca and Stille (1996) and Tricca (1997).

Fig. 4.23 illustrates the relationship between Sr concentration and Sr isotopic composition in Rhine water and pit water. The $^{87}Sr/^{86}Sr$ ratios of the waste water from the potassium mines are around 0.7095, while the concentration is 31.4 mg/l. The isotopic composition and concentration of Sr in the non-contaminated Rhine water are 0.7084 and 0.38 mg/kg, respectively. Rhine water below the introduction of this contaminated water shows a significantly higher Sr isotopic composition of 0.7086-0.7087 and concentrations of 0.46-0.52 mg/kg (Fessenheim). All sample points lie on a mixing line. The samples from Fessenheim are the results of mixing uncontaminated Rhine water and rinse water from the pits. Using these data and the equation system for the mixing of two components (Faure 1986; page 141), one can calculate that about 6 m^3 of pit water per second flows into the Rhine. As the main and trace element concentrations in the contaminating tributary are known at this point, then the rate of flow for the contaminating elements can be calculated (Tricca 1997; Tricca and Stille 1996).

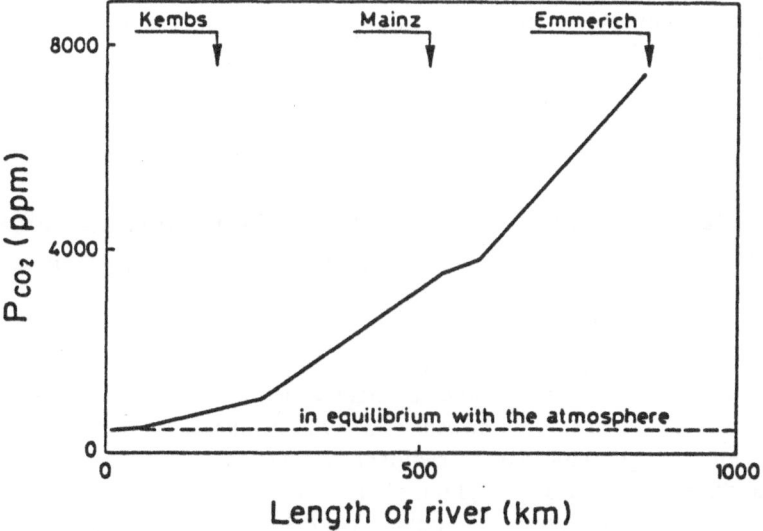

Fig. 4.25. The evolution of dissolved CO_2 content in the Rhine water. (Bühl et al. 1991)

This study clearly shows that the Sr isotope method is a valuable tool in the investigation of hydrogeochemical exchange processes between different water systems. In the special case of the potassium mines, chemical exchange processes between groundwater and pit rinse water can be described and quantified.

4.3.3 Carbon Isotopes

$\delta^{13}C$ values decrease down river from -4 to -9 ‰ PDB (see Sect. 1.1) and reach a minimal value of -10.7 ‰ below Mainz (Fig. 4.24). Almost all tributaries show lower $\delta^{13}C$ values than the Rhine itself. The carbon isotope compositions are strongly influenced by the input of organic carbon from local tributary rivers. The $\delta^{13}C$ value of organic carbon is about -25 ‰, whereas that of carbon from dissolved carbonate is about 0 ‰. By applying these end members one can calculate that the proportion of organic carbon in the Rhine climbs from 18% in the Alpine regions to 43% close to the entrance of the river Main into the Rhine. 60% of the carbon near Mainz originates from organic sources. The tributaries, which are rich in organic carbon significantly influence the carbon budget of the Rhine. Over fertilization and resultant enhanced biologic productivity are responsible for such high concentrations of organic material in our waterways.

Such over production of organic, microbially degradable materials has the consequence that there is a strong tendency for oxygen to be removed and CO_2 to be produced. This can be seen in CO_2 analyses of Rhine water (Fig. 4.25; Bühl et al. 1991). Over a distance of 500 km the amount of CO_2 in water doubles because of CO_2 contributions from tributaries as can be seen from isotopic analyses of the carbon in these tributaries.

4.4 Contamination of Flowing Water and Groundwater after the Reactor Accident at Chernobyl Shown Using Released Radionuclides.

The most serious accident at a nuclear plant which led to the release of nuclear energy in peace time occurred on April 26, 1986 in Chernobyl (Ukraine). During this accident large quantities of radioactive materials were released into the atmosphere and dispersed throughout almost the whole of Europe. This input of large amounts of mother and daughter isotopes into the biosphere had the consequence that both the soils and the hydrosphere experienced serious contamination. These chemical and radioactive contaminants are not only dangerous for humans, animals and plants but also weaken the microorganisms which help to regulate the soil and the groundwater.

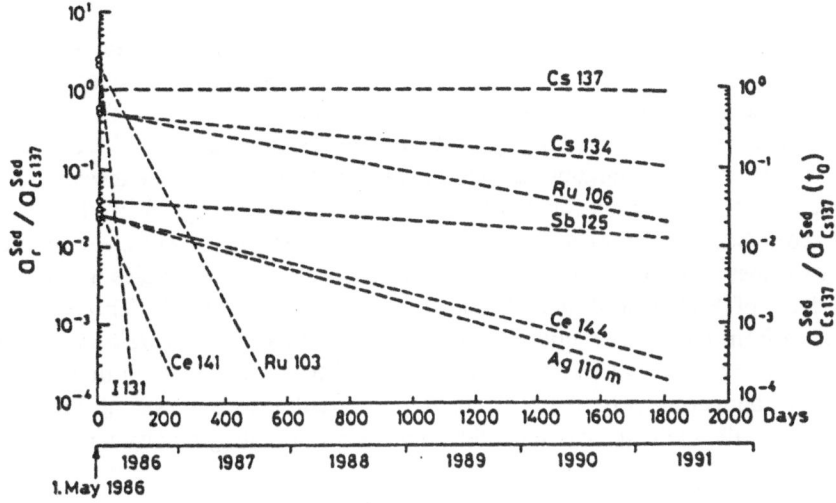

Fig. 4.26. Activities of various radionuclides normalized to ^{137}Cs shown as a function of time. (Mundschenk 1992)

The Chernobyl accident made it possible to investigate the transport mechanisms of these contaminants in the rivers and their infiltration and migration into soils and groundwaters. In the studies we will cite, the following radionuclides were looked at in particular: ^{103}Ru, ^{106}Ru, ^{131}I, ^{132}Te, ^{141}Ce, ^{144}Ce, ^{125}Sb, ^{134}Cs and ^{137}Cs (Cesium). Most of these radionuclides have already decayed further. In Fig. 4.26, activity quotients of various radionuclides are displayed as a function of time, normalized to ^{137}Cs. The reference time point t_0 is May 1, 1986. This figure allows us to reconstruct that by the beginning of 1991, alongside the relatively long lived ^{137}Cs (half life = 30.2 years), only the nuclides ^{134}Cs, ^{106}Ru and ^{125}Sb were still present in sufficient quantity to be detected. The nuclide quotients assumed for the reference point t_0 represent those, which have been determined from sediments of the Rhine. Mundschenk (1992) investigated the large scale consequences of the reactor accident on the water, suspended material and sediments of German river systems. As a guide, he took the example of the long lived nuclide ^{137}Cs. Fig. 4.27 offers some insight into the behavior of ^{137}Cs content in unfiltered water samples (a_r^W) from the Rhine and the Mosel between 1986 and 1991. We can see that by 1991, concentrations had practically fallen back to the low values of before the reactor accident.

The filtered radionuclide enriched suspended phases, show similar concentration trends (a_r^{susp}) to those of unfiltered water, although the concentrations are far higher (Fig. 4.28). Mundschenk was able to calculate from the nuclide content of the unfiltered bulk sample a_r^W, the proportion found in suspension and the nuclide content of this suspended phase (a_r^{susp}), and that for example, 60-70 % of the cesium isotopes ^{134}Cs und ^{137}Cs exist in dissolved form.

The contaminated phases in suspension can be deposited in calm waters, and can therefore lead to contamination of the river bottom sediments themselves. The river sediments investigated show Cs concentrations, which are similar to those of the suspended phases. However, there can be large variations in concentration from site to site (Fig. 4.29). Mundschenk (1992) related these variations to grain size effects (Fig. 4.30). From Fig. 5.30 we can tell that not only ^{137}Cs but also ^{106}Ru and ^{144}Ce concentrations are strongly dependent on grain size distribution in the sediment. Thus, ^{144}Ce could only be detected in the finest grain size fraction of the sediment. Fine grained sediments are obviously more enriched in these radionuclides than coarse grained sediments. Other important studies in this vane are those of Waber et al. (1987) and von Gunten et al. (1988). These authors profited from the permanently installed system of investigation wells, which are located near the river Glatt (Switzerland); they have already been discussed in another connection (Fig. 4.12). Here it was possible to observe the migration behavior of the radionuclides released from Chernobyl into groundwater systems.

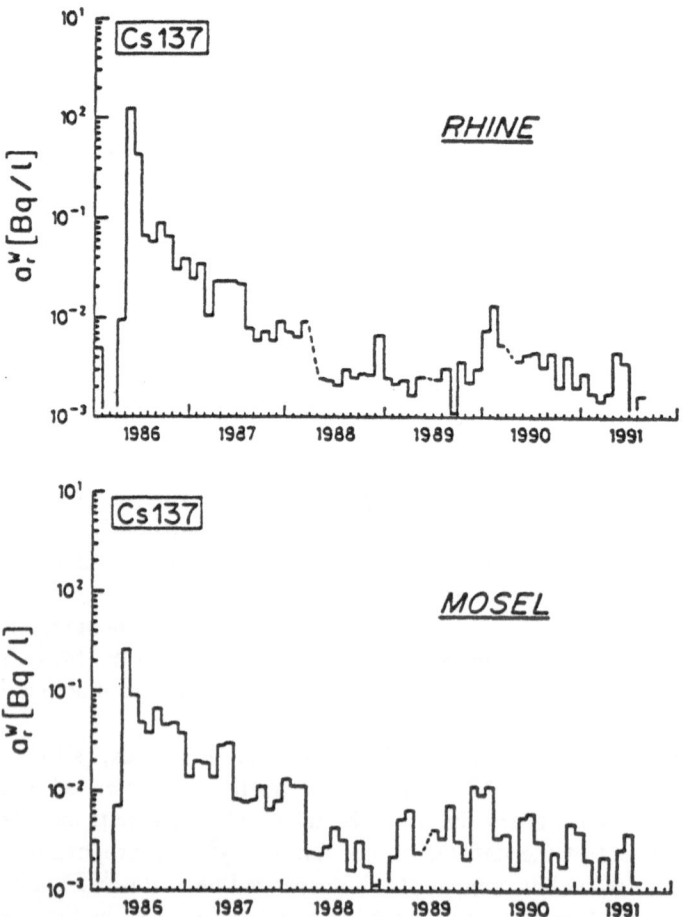

Fig. 4.27. ^{137}Cs activity in unfiltered water samples from the Rhine and Mosel rivers between 1986 and 1991. (Mundschenk 1992)

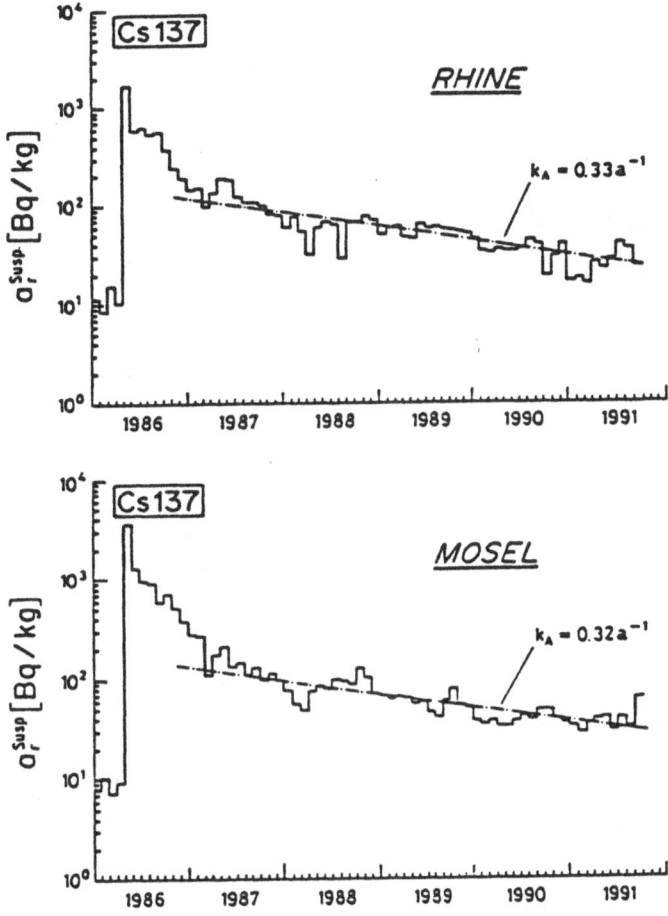

Fig. 4.28. ^{137}Cs activity in the suspended load from the Rhine and Mosel rivers between 1986 and 1991. (Mundschenk 1992)

They determined the contents of the radionuclides ^{103}Ru, ^{131}I, ^{132}Te, ^{134}Cs and ^{137}Cs in river water and in groundwater. Sample collection took place from just one to four weeks after the accident.

Tables 4.7 und 4.8 display the activities (Bq/l) of radionuclides in river water and groundwaters. D stands for the dissolved phase and P for particulates/colloids. From this we can observe that the particles/colloids transported less than 25% of the radioactivity into river water. The main activity could be detected in filtered water. This means that the radionuclides investigated here exist in molecular solution or were adsorbed onto very small colloids ($<0.05\mu$m). Fig. 4.31 shows the distribution of activity in dissolved and particulate phases in river water.

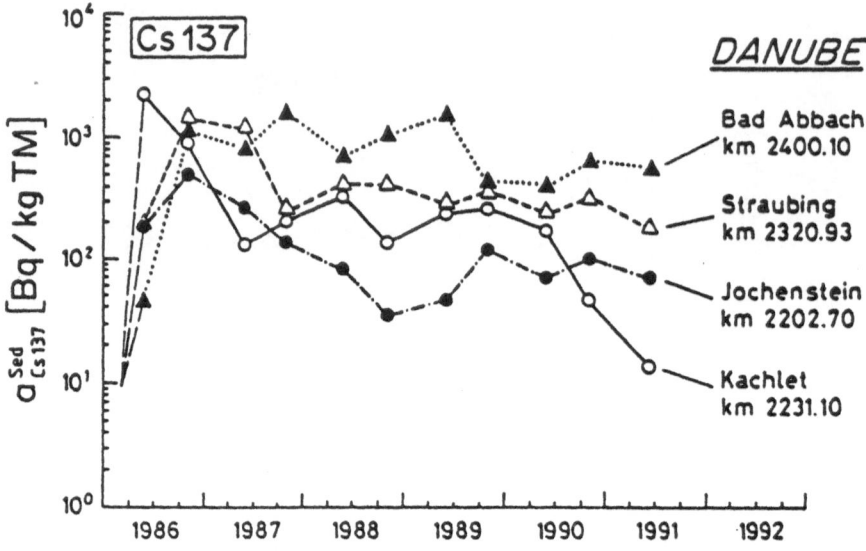

Fig. 4.29. ^{137}Cs activity in Donau river sediments between 1986 and 1991. (Mundschenk 1992)

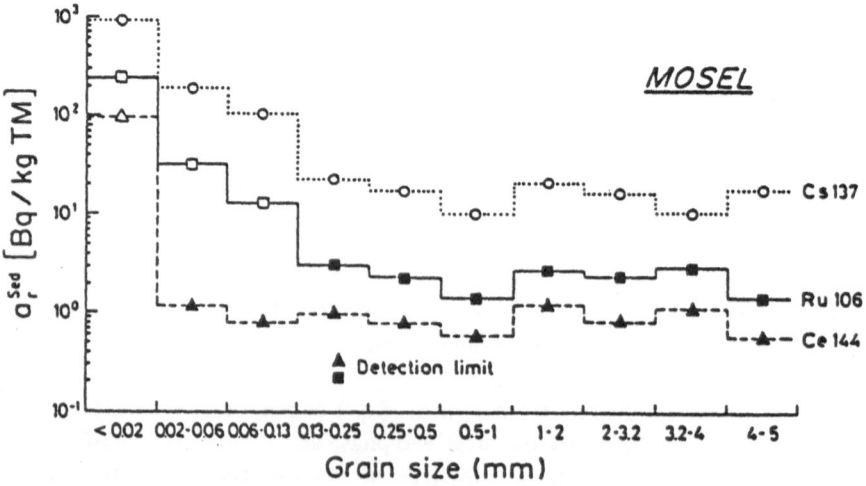

Fig. 4.30. ^{137}Cs, ^{106}Ru and ^{144}Ce activities in sediments as a function of grain size. (Mundschenk 1992)

Table 4.7. Radioactivity (BqL^{-1}; errors ±10%) on particles and filtrated river water. (after von Gunten et al. 1988)

Date 1986	Filtration (sizes in µm)	Nuclides I-131	Te-132	Ru-103	Cs-134	Cs-137
5.2.	D, folded filter	22.2	20.3	1.9	0.9	1.2
	P, folded filter	2.2	3.1	0.7	0.2	0.25
5.4.	D, folded filter	7.0	10.9	1.1	0.4	0.8
	P, folded filter	0.9	1.0	0.2	0.04	0.08
5.5.	D, <0.05	8.1	9.0	0.9	<0.06	0.6
	P, >0.45	1.0	2.1	0.6	0.2	0.2
	P, 0.2 - 0.45	0.04	0.2	0.05	0.01	0.01
	P, 0.05 - 0.2	0.03	0.1	0.02	nd	0.01
5.12.	D, <0.05	10.7	7.0	1.6	0.7	1.1
	P, >0.45	0.4	1.1	0.1	nd	0.1
	P, 0.2 - 0.45	0.3	0.06	0.2	nd	0.05
	P, 0.05 - 0.2	0.2	nd	0.04	nd	0.06
5.20.	D, <0.05	6.4	nd	1.1	nd	0.4
	P, >0.45	0.3	<1.7	0.06	0.04	0.04
	P, 0.2 - 0.45	0.04	<0.5	0.02	<0.001	0.01
	P, 0.05 - 0.2	0.03	nd	0.01	nd	0.01

D: dissolved load; P: suspended load/particles; nd: not detected

Table 4.8. Radioactivity (BqL^{-1}; errors±~10%) on colloids and in filtrated groundwater. Activities normalized to April 30th, 1986. (after von Gunten et al. 1988)

Date May 1986	Distance of of well from river (meters)	Filtration (µm)	Nuclides I-131	Te-132	Ru-103	Cs-134,137
5	2.5	D, <0.05	7.4	2.9	0.8	nd
		P, >0.05*	0.04	0.05	0.02	nd
	5	D, <0.05	12.8	3.3	0.9	nd
		P, >0.05*	0.02	0.05	nd	nd
13	2.5	D, <0.05	12.8	nd	1.2	nd
		P, >0.05*	0.1	<0.09	<0.002	<0.002
	5	D, <0.05	8.7	nd	0.9	nd
		P, >0.05*	<0.004	<0.09	<0.001	<0.002
	4.5	D, <0.05	<0.1	nd	<0.04	<0.03
	deep well	P, >0.05*	<0.003	<0.08	<0.001	<0.001

D:dissolved load; P:particulates/colloids;*Sum of activity on 0.45, 0.2, 0.05 µm filters

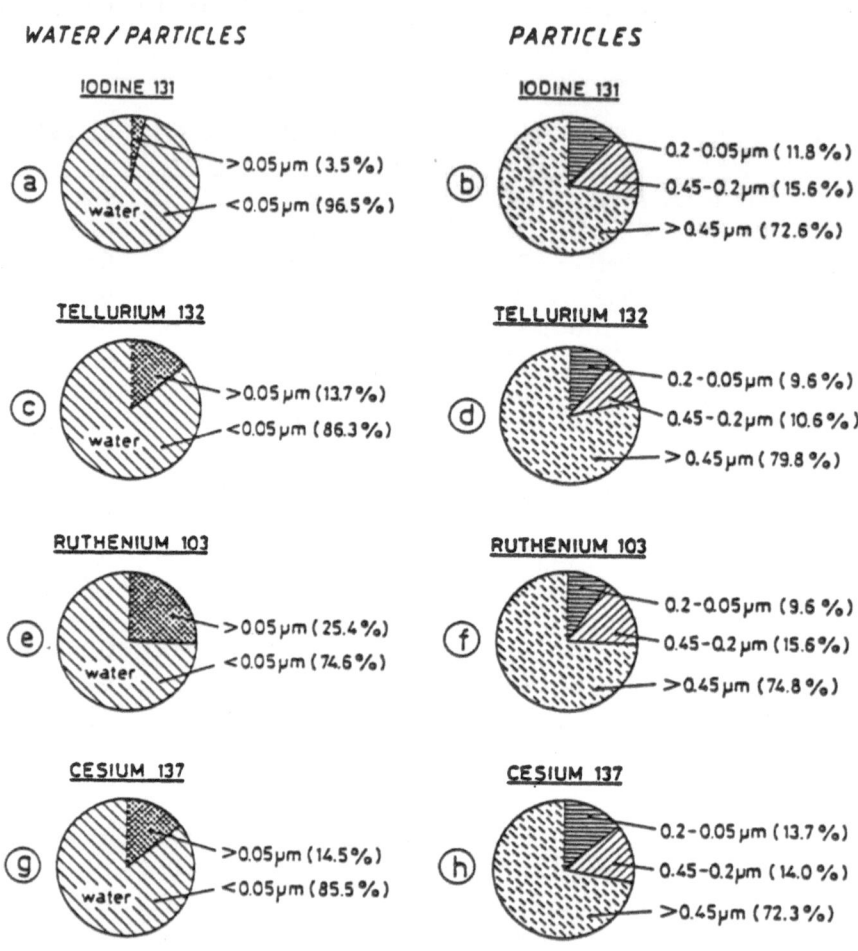

WATER / PARTICLES

PARTICLES

IODINE 131

(a)
water
>0.05 μm (3.5%)
<0.05 μm (96.5%)

IODINE 131

(b)
0.2-0.05μm (11.8%)
0.45-0.2μm (15.6%)
>0.45μm (72.6%)

TELLURIUM 132

(c)
water
>0.05 μm (13.7%)
<0.05 μm (86.3%)

TELLURIUM 132

(d)
0.2-0.05μm (9.6%)
0.45-0.2μm (10.6%)
>0.45μm (79.8%)

RUTHENIUM 103

(e)
water
>0.05 μm (25.4%)
<0.05 μm (74.6%)

RUTHENIUM 103

(f)
0.2-0.05μm (9.6%)
0.45-0.2μm (15.6%)
>0.45μm (74.8%)

CESIUM 137

(g)
water
>0.05μm (14.5%)
<0.05μm (85.5%)

CESIUM 137

(h)
0.2-0.05μm (13.7%)
0.45-0.2μm (14.0%)
>0.45μm (72.3%)

Fig. 4.31. Activity distribution in river water (Glatt, Switzerland) between the dissolved (<0.05 μm) and the particulate fraction (>0.05μm) and among the different size fractions of the particles. (Waber et al. 1987)

The groundwater samples come from sources that were 2.5 m and 5 m beneath the river bed. Groundwater samples that came from greater depths showed no activity and were thus not contaminated by the fall out after Chernobyl (Table 4.8). As with river water, the filtered groundwater shows strongly enhanced radioactivities compared with the material in suspension/colloids. Less than 2.5% of the investigated radionuclides could be detected in the colloidal phases. The contamination of the groundwater by these elements must take place mainly by way of dissolved phases.

4.5 Radionuclides as Tracers in Sedimentology

Atomic bomb trials carried out between 1945 and 1972 in the atmosphere had the consequences that by 1955 significant quantities of radionuclides were already reaching the floors of lakes and rivers. The largest activity peak is usually associated with the period shortly before the first wide reaching test ban. During this time, several countries made 'good' use of their last opportunity for unrestricted testing. The long lived Cesium isotope (^{137}Cs) can still be detected in river sediments today. As we saw in the preceeding section, the reactor accident of Chernobyl was likewise characterized by an increase in the activity of such radionuclides, thus creating a second time marker in the sediments. This is demonstrated in the investigations of Mundschenk (1992) working on German river sediments. With the help of the activity of the marker cesium nuclide (^{137}Cs), analyzed from sediment cores taken from a typical small port (Fig. 4.32), both events can be distinguished. If we can find both time markers, we can therefore obtain some idea of the sedimentation rate.

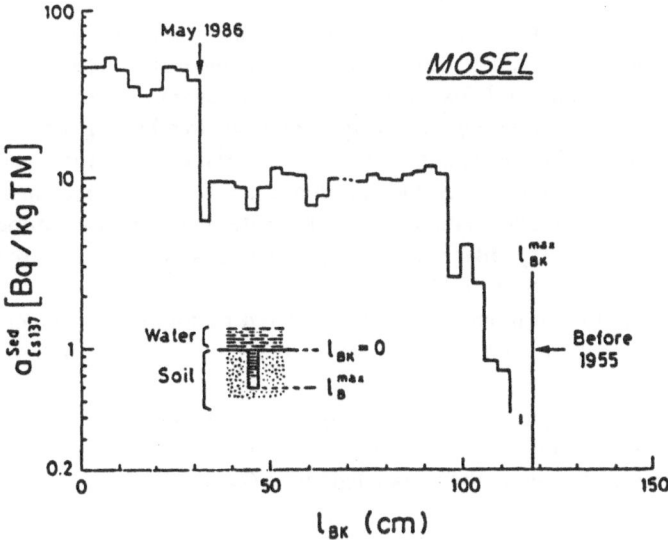

Fig. 4.32. Concentration profile of a sediment core from a harbor at the Moselle river (Germany) using ^{137}Cs as tracer. (Mundschenk 1992)

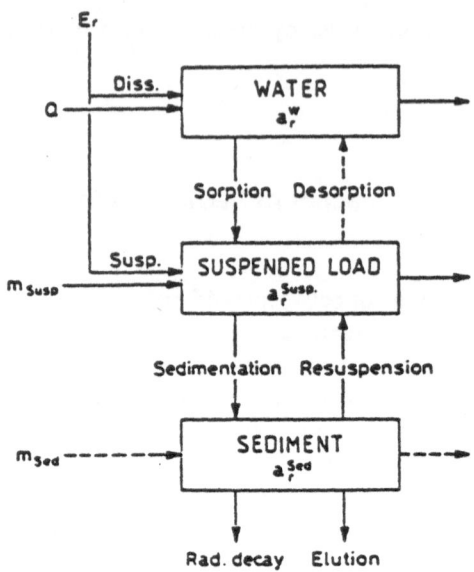

Fig. 4.33. Transfer of radionuclides from the water and suspended load up to sedimentation. (Mundschenk and Tolksdorf 1988)

Mundschenk und Tolksdorf (1988) developed a model that describes the transfer of radionuclides from the water or from material in suspension in water up to sedimentation (Fig. 4.33). The toxic radionuclides are brought into the rivers either in dissolved or suspended form. Their transport up to the time of sedimentation depends upon various physico-chemical factors. Isotopic exchange at the phase boundaries between material in suspension and water as well as sorption and agglomeration processes quickly leads to a binding of the introduced radionuclides and the suspended material. Contamination of the river sediments takes place as a result of the sedimentation of the radionuclide enriched suspended particles. The radioactive contamination of a water body and its sediments can be controlled by two very different input mechanisms according to the model of Mundschenk und Tolksdorf (1988) (see Fig. 4.34):

1. an almost continual input ($E_{v,r}$) of long lived radionuclides, which leads to a constant source of contamination due to the erosion of a contaminated soil ($a_{v,r}^{susp}$).

2. a variable input source (E_r), for example from nuclear power plants. This sort of input, which is highly time constrained, leads to a momentary increase in the content of radionuclides in the suspended phase (Δa_r^{susp}).

In this model it is assumed that no resuspension takes place. The rate of sedimentation:

$$v_{sed} = \frac{\Delta d}{\Delta T_s} \tag{I}$$

ΔTs corresponds to the time period during which sedimentation takes place leading to an increase in the sediment pile thickness equal to Δd (Fig. 4.34). In bed i, the nuclide content present: $a_{r,i}{}^{sed}$ (Bq/kg dry sample) is considered to be the result of the influence of the two independent input mechanisms (see above):

$$a_{r,i}^{sed} = \Delta a_{r,i}^{susp} + a_{v,r,i}^{susp} \tag{II}$$

The average nuclide content at time T_p in the sediment sample $a_r{}^{sed}(T_p)$ can be calculated as follows after consideration of the appropriate radioactive decay constant (λ):

$$a_r^{Sed}(T_p) =$$

$$\frac{\Delta d_1 a_{r,1}^{sed} \exp\left[-\lambda\left(T_p - T_{s.1}\right)\right] + \ldots \Delta d_n a_{r.n}^{sed} \exp\left[-\lambda\left(T_p - T_{s.n}\right)\right]}{\sum\limits_{i=1}^{n} \Delta d_i} \tag{III}$$

Assuming that the sediment bed thicknesses Δdi are the same throughout period ΔTs then the sedimentation rate v_{sed} must also be constant through time. Therefore:

$$\Delta d_1 = \Delta d_2 = \ldots\ldots\ldots \Delta d_i = \Delta d$$

$$\sum\limits_{i=1}^{n} \Delta d_i = d$$

$$T_p - T_{s.i} = i\Delta T_s = \frac{i\Delta d}{v_{sed}} \tag{IV}$$

We can rearrange equation (III), by taking into account equations (II) and (IV) as follows:

$$a_r^{sed}(T_p) = \frac{\Delta d}{d}\left[\sum\limits_{i=1}^{n} \Delta a_{r,i}^{susp} \exp\left(-\lambda\frac{i\Delta d}{v_{sed}}\right) + \sum\limits_{i=1}^{n} a_{v,r,i}^{susp} \exp\left(-\lambda\frac{i\Delta d}{v_{sed}}\right)\right] \tag{V}$$

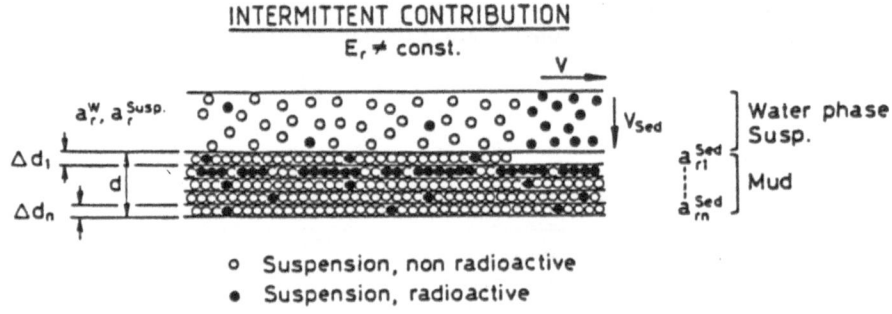

Fig. 4.34. The radioactive contamination of a water body and its sediments (bed i) can be controlled by two very different input mechanisms according to a model of Mundschenk and Tolksdorf (1988).

For long-lived radionuclides such as Cesium ^{137}Cs :

$$\lambda \longrightarrow 0$$
$$E_{v,r} \approx konst$$

Equation (V) can be rearranged as follows:

$$a_r^{sed}(T_p) = \frac{\Delta d}{d}\left[\sum_{i=1}^{n}\Delta a_{r.i}^{susp} + na_{v.r.i}^{susp}\right]$$

(VI)

These equations allow us to estimate the concentrations of long lived radionuclides (such as ^{137}Cs) in the sediment, which has been contaminated by intermittent input (E_r) of radioactive material. The radionuclide determinations that were carried out by Mundschenk und Tolksdorf (1988) show very good agreement between model calculations and directly determined values.

4.6 References

Ault UW, Senechal RG, Erlebach E (1970) Isotopic composition as a natural tracer of lead in the environment. Environmental Science and Technology, 4: 305-317

Bühl D, Neuser RD, Richter DK, Riedel D, Roberts B, Strauss H and Veizer J (1991) Nature and Nurture: environmental isotope story of the River Rhine. Naturwissenschaften 78: 337-346

Boutron CF, Patterson C (1986) Lead concentration changes in Antarctic ice during the Wisconsin/Holocene transition. Nature, 323: 222-225

Elbaz-Poulichet F, Holliger P Huang WW, Martin JM (1984) Lead cycling in estuaries, illustrated by the Gironde estuary, France. Nature, 308: 405-414

Elbaz-Poulichet F, Holliger P, Martin JM, Petit D (1986) Stable lead isotope ratios in major french rivers and estuaries. The Science of the Total Environment, 54:61-76

Erel Y, Patterson CC, Scott MJ, Morgan JJ (1990) Transport of industrial lead in snow through soil to stream water and groundwater. Chem Geol, 85:383-392

Esser BK, Turekian EK (1993) Anthropogenic osmium in coastal deposits. Environ Sci Tech, 27: 2719-2724

Flegal AR, Nriagu JO, Niemeyer S, Coale KH (1989) Isotopic tracers of lead contamination in the Great Lakes. Nature, 339: 455-457

Hoehn E, von Gunten HR (1989) Radon in Groundwater: A tool to assess infiltration from surface waters to aquifers. Water Resources Res, 25:1795-1803

Horn P, Michler G, Todt W (1987) Die anthropogene Blei Belastung im Raum München, ermittelt aus Pb-Isotopenmessungen von Wasser- und Sedimentproben. Mitt der Geograph Gesellschaft in München, 72: 105-117

Krishnaswami S, Graustein WC, Turekian KK (1982) Radium, thorium, and radioactive lead isotopes in groundwaters: Application to the in-situ determination of adsorption-desorption rate constants and retardation factors. Water Resour Res, 18/6: 1633-1675

Kummert R, Stumm W (1989) Gewässer als Ökosysteme. Verlag der Fachvereine, B.G. Teubner Verlag, Stuttgart

Mundschenk H (1985) Über Auswirkungen von radioaktiven Ableitungen aus Kernkraftwerken in staugeregelten Flüssen am Beispiel des Neckars. Teil II: Radiometrische Messungen und radiologische Bewertung. DGM, 29: 134-144

Munschenk H (1992) Über Nachwirkungen des Reaktorunfalls in Tschernobyl im Bereich der "alten" Bundeswasserstrassen. DGM, 36: 7-19

Mundschenk H, Tolksdorf, W (1988) Methodische Untersuchungen zur Sedimentation mit Hilfe radioaktiver Leitstoffe am Beispiel eines Kleinhafens am Mittelrhein. Deutsche Gewässerkundliche Mitteilungen, 32: 110-119

Murozimi M, Chow TJ, Patterson C (1969) Chemical concentrations of pollutant lead aerosols, terrestrial dusts and sea salts in Greenland and Antarctic snow strata. Geochim Cosmochim Acta, 33:1247-1294

Ravizza GE, Bothner, MH (1996) Osmium isotopes as tracers of anthropgenic metals in sediments from Massachusetts and Cape Cod Bays. Geochim Cosmochim Acta 60, 15: 2753-2763.

Siegenthaler U, Schotterer U (1977) Hydrologische Anwendungen von Isotopenmessungen in der Schweiz Gas, Wasser, Abwasser, 57: 501-506

Sigg L (1985) Chemical processes in lakes. Wiley, New York

Sigg L, Stumm W (1989) Aquatische Chemie. Verlag der Fachvereine, B.G. Teubner Verlag, Stuttgart

Steinmann M, Stille P (1997) Rare earth element behavior and Pb, Sr, Nd isotope systematics in a heavy metal contaminated soil. Applied Geochem (in press)

Stichler WP, Maloszewski P, Moser H (1986) Modelling of river water infiltration using $\delta^{18}O$-data. J Hydrol, 83: 355-365

Sturges WT, Barrie LA (1987) Lead 206/207 isotope ratios in the atmosphere of North America as tracers of US and Canadian emissions. Nature, 329: 144-146

Tricca A (1997) Transport mechanisms of trace elements in surface and ground water: Sr, Nd, U and Rare Earth Element evidence. PhD Thesis, Uni Strasbourg

Tricca A, Stille P (1996) Rare earth elements and Sr isotope determinations on Alsatian River and Groundwaters. VM Goldschmidt Conf, J of Conf Abstracts,1,626

Van der Weijden CH, Middelburg JJ (1989) Hydrogeochemistry od the River Rhine: long term and seasonal variability, elemental budgets, base levels and pollution. Wat Res 23: 1247-1266

Von Gunten HR, Waber U, Krähenbühl U (1988) The reactor accident at Chernobyl: a possibility to test colloid-controlled transport of radionuclides in a shallow aquifer. Journal of Contaminant Hydrology, 2: 237-247

Waber U, von Gunten HR, Krähenbühl U (1987) Radiochimica Acta, 41: 191-198

Wolff EW, Peel DA (1985) The record of global pollution in polar snow and ice. Nature 313: 535-540

Zobrist J (1983) Die Belastung der Gewässer mit Schadstoffen aus Abwässern und Niederschlägen. Gaz-Eaux-Eaux usées, 3: 123-131

5 Isotopic Composition of Seawater Past and Present (Sr, Nd, Pb, Os, Ce)

Most material that has undergone weathering on the continent ends up in the sea either in suspended or in dissolved form, where it is subsequently deposited either as detritus or as a chemical precipitate, respectively. Therefore, seawater represents the most important environment for the formation of sediments. Isotopic information about the marine environment today can help towards making material mass balances for the oceans as well as shedding light on circulation patterns. Isotopic information about past oceans can help us not only to date sediments by establishing secular trends in isotopic ratios through time but also to constrain the various parameters which may influence an isotopic ratio at any one time, such as tectonics, weathering rates, ocean spreading rates, cosmogenic input, anthropogenic contamination or biological evolution. An exciting aspect of this research is how various isotopic systems, each of which may reveal different information to us, can be brought together in conjunction with more traditional geological techniques, to bear on any one specific question.

Research in this area has been made easier by the efforts of the ODP (Ocean Drilling Project). Seawater has been sampled on numerous occasions at geographically various locations and as a result, the isotopic and chemical homogeneity of the oceans is well known for the major and trace elements. Some elements possess isotopic ratios, which are more or less identical in open seawater worldwide (e.g. Sr, S), whereas others vary so much that they can be used as tracers for circulation patterns (e.g. O, Nd, Pb, Ce). Biologically related isotopic fractionation processes only affect the stable isotopes, (e.g. C, O and S). Therefore, the isotopic homogeneity in seawater of the much heavier radiogenic elements is related to how quickly an element can be removed from seawater relative to the size of the whole reservoir only. This useful parameter, reservoir size/flux, is referred to as the 'residence time' and its significance will be explained more fully using Sr and Nd as examples. For information about past oceans, it is necessary to analyze sedimentary minerals that formed in isotopic equilibrium with coeval seawater, i.e. authigenic minerals. The oldest sediments that have been drilled offshore are from the Pacific Ocean and are no more than 160 million years old. If well preserved, these minerals can yield important information back to the Jurassic Period. For information about even more ancient oceans we must turn to well preserved sediments on land. Relating the composition of now lithified rocks to the composition of a paleo-ocean brings many potential problems that we will discuss later using case studies. Please note that this chapter deals primarily with the radiogenic isotopes. There are several good teaching texts already available on the subject of stable isotopes in the marine environment, e.g. Hoefs (1980) and Arthur and Anderson (1983).

5.1 Sr Isotopic Composition of Seawater: Tectonic Proxy and Stratigraphic Tool

Strontium is present in seawater as a trace element with a concentration of 7-8 mg/L. Due to its very long residence time (about >3 Ma) relative to the ocean mixing time (about 1500 years: Broeker and Peng, 1982), the isotopic composition of Sr does not change detectably from place to place in the open ocean: $^{87}Sr/^{86}Sr$ = 0.709165 +/- 0.000020, (this value has been normalized to the commonly reported value for the standard NBS (National Bureau of Standards) 987 of 0.710250). Only in restricted basins, partly closed to seawater, are we likely to find any significant deviation from the seawater ratio. For example, seawater and therefore also shells and carbonate cement from sabkha environments or restricted seaways such as the Gulf of Bothnia in the Baltic Sea may not possess a seawater isotopic signature (Aberg and Wickman, 1987). Significantly anomalous isotopic ratios may also be observed in areas of high hydrothermal activity, for example, along the mid-ocean ridge (Albarède et al., 1981). Investigations of hydrothermal vents of the "East Pacific Rise" near to the "Galapagos Spreading Center" make it clear that the seawater Sr isotopic signature there has been influenced by mantle Sr. Fig. 5.1 helps to illustrate these exchange mechanisms.

The three samples of hydrothermal waters (open circles) show variable $^{87}Sr/^{86}Sr$ and Mg/Sr ratios which are positively correlated with each other. Pure, non-hydrothermally contaminated seawater lies on the mixing line with the characteristic ratio of 0.70916 and with high Mg/Sr ratios of about 600. Mg in the mixing system is likely to come from seawater only. A non-seawater contaminated hydrothermal system has a Mg/Sr of about zero and a $^{87}Sr/^{86}Sr$ value similar to the basalts, i.e. about 0.703.

It is interesting to note that there is a direct correlation between the temperature of the investigated waters and their Sr isotope composition. This correlation allows us to reconstruct an initial temperature of about 340 +/- 30°C for the hydrothermal fluid phase before any exchange with seawater.

Because of its high absolute mass, strontium does not experience any isotopic fractionation or 'vital effect' during shell formation as does carbon. Sr is also not

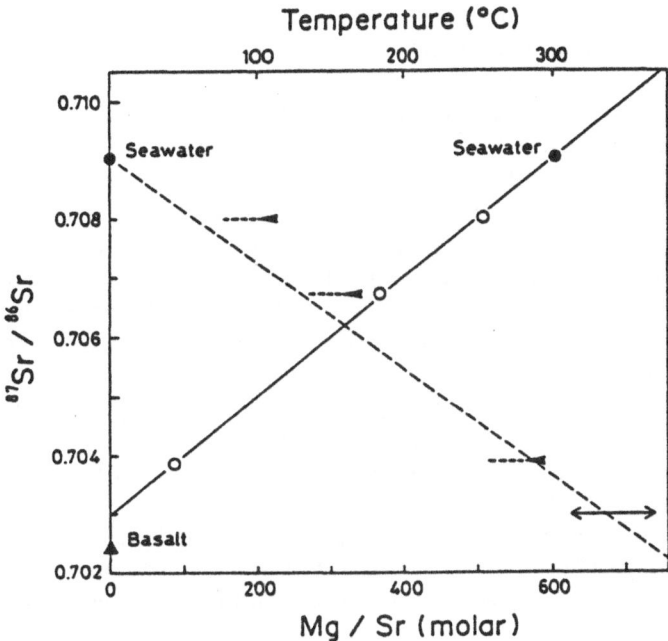

Fig. 5.1. The seawater $^{87}Sr/^{86}Sr$ and Mg/Sr ratios near the 'Galapagos spreading center' (Pacific) show clearly that local seawater has been influenced by mantle Sr. If all the Mg in this mixing system derives from seawater it can be assumed that the hydrothermal fluid system uncontaminated with seawater must show a similar $^{87}Sr/^{86}Sr$ ratio to basalt, i.e. 0.703. A simple correlation between water chemistry and temperature (arrows) allows the temperature of the hydrothermal fluid phase to be determined at 340 °C. (Albarède et al. 1981)

fractionated during reduction or oxidation or by any other kinetic effects during chemical reactions at the Earth's surface.

Thus, the isotopic ratio of Sr in the oceans ought to reflect the relative importance of the various sources of Sr only. This principle has formed the basis for much of the research, not just on Sr isotopes in the marine environment but for all the heavy radiogenic elements.

5.1.1 Microstratigraphy

Wickham (1947) was the first to suggest the use of the Sr isotope ratio as a stratigraphic tool. He proposed that seawater $^{87}Sr/^{86}Sr$ had increased predictably through time due to the sluggish decay of ^{87}Rb to ^{87}Sr. Although this is true on the time-scale of the whole age of the Earth (cf. Fig. 5.24), initial research showed quickly that this was not the case during the Phanerozoic and that the seawater Sr isotope ratio had fluctuated significantly in the geologic past depending on which source of Sr was more dominant at any one particular time.

The most important sources of dissolved Sr in the sea are considered to be 1) products of erosion carried by rivers and the wind, 2) ancient and recent marine carbonate dissolution on the seafloor, and 3) exchange with hydrothermal sources on the ocean floor. Peterman et al. (1970) showed in an isotope study on carbonate fossil fragments that the Sr isotopic ratio of seawater underwent strong fluctuations during the Phanerozoic era. The authors explained these fluctuations by invoking variable input of crustal Sr with high $^{87}Sr/^{86}Sr$ ratios. Increases in seawater $^{87}Sr/^{86}Sr$ would not necessarily correlate with enhanced continental weathering but would also reflect the Sr isotopic ratios (ages and initial Rb/Sr ratios) of the crustal rocks undergoing weathering. Veizer and Compston (1974) came to similar conclusions. Clauer (1976) considered that "sea-floor spreading" and thus the opening of the ocean basins at various times had also strongly influenced seawater $^{87}Sr/^{86}Sr$.

Fig. 5.2. shows the results of 786 measurements of bulk marine carbonates of Phanerozoic age (Burke et al. 1982). The broad temporal trends in seawater $^{87}Sr/^{86}Sr$ ratio implied in these data have been confirmed by every study since.

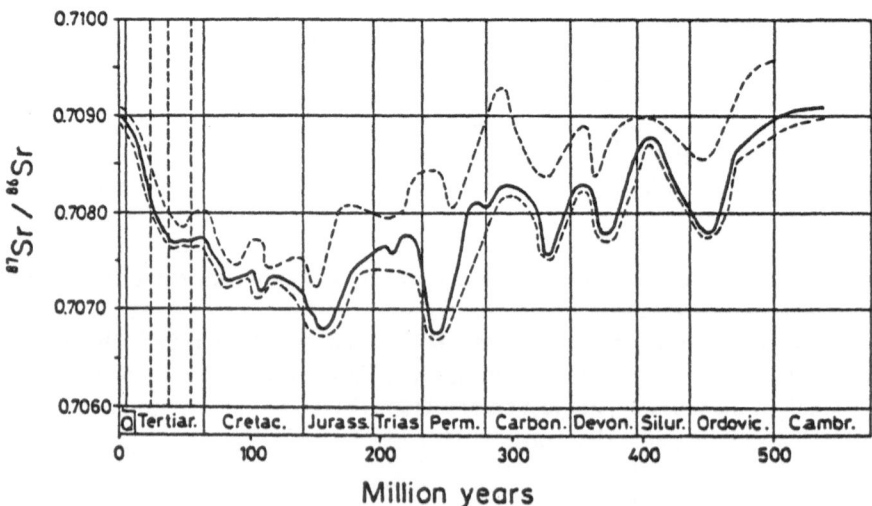

Fig. 5.2. Temporal variations of the Sr isotope ratio in seawater based on 786 isotopic analyses of marine carbonate rocks. The dashed lines show the extent of 95% of the data points. (Burke et al., 1982)

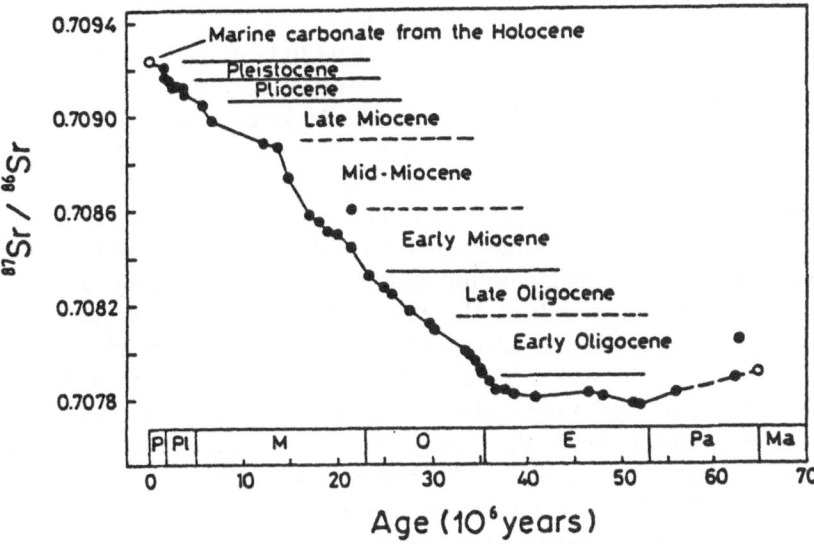

Fig. 5.3. The precise $^{87}Sr/^{86}Sr$ isotopic analyses of DePaolo and Ingram (1985) on stratigraphically well constrained marine carbonates allowed the reconstruction of a single clear trend in Tertiary seawater isotopic variation for the first time.

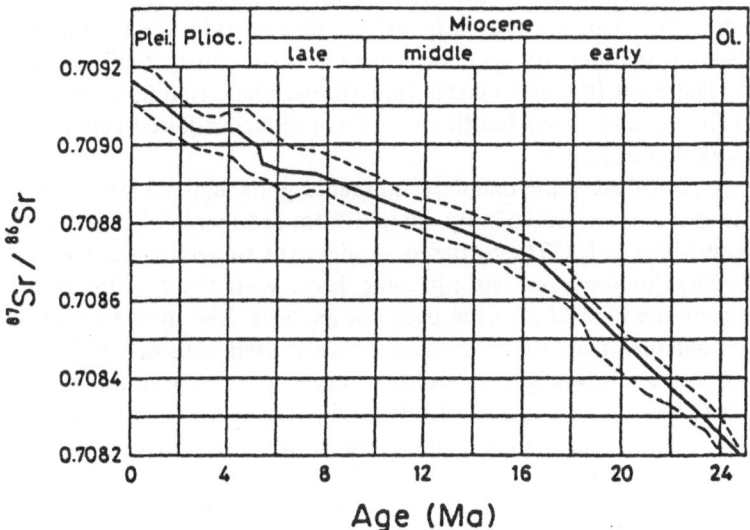

Fig. 5.4. 261 Sr isotopic analyses on planktonic foraminifera allowed Hodell et al. (1991) to constrain Sr isotopic evolution in the Neogene ocean very precisely. Such secular trends permit isotope microstratigraphy. Subsequent studies replicated these results within the error permitted by the analytic technique.

The consistency shown in samples from any one stratigraphic level show that it is a normal situation for the marine environment to be isotopically homogeneous with respect to Sr even over geological time-scales confirming that Sr has a long residence time in seawater. The scatter in the Burke et al. curve is due to sample alteration and errors in stratigraphic correlation; we will return to the significance of this scatter and how much of it can be removed successfully later.

Where samples are well preserved, data scatter can be minimalized. Very precise $^{87}Sr/^{86}Sr$ ratios of stratigraphically well constrained marine carbonates (bivalve pieces) and planktonic foraminifera were published by DePaolo and Ingram (1985) and Hodell et al. (1991). Such studies have made detailed reconstructions possible of of seawater $^{87}Sr/^{86}Sr$ evolution in the Neogene oceans (Fig. 5.3, 5.4). Marine carbonates could be dated and correlated precisely with the help of the Sr isotope system. Since 1991, further studies have yielded similar results. Paytan et al. (1993) also reproduced the same trend through the Neogene using biogenic barite particles. Interestingly enough, their results were, if anything, slightly lower than those from foraminifera and marine carbonate. Such deviations from the standard curve commonly lie within the error provided by analytic precision and reproducibility and so remain controversial (e.g. see Dia et al., 1992 and follow-up in Henderson et al., 1994).

Stille et al. (1994) discuss the stratigraphic potential of Sr isotopes in their study of the world's largest Neogene phosphate deposits: North Carolina and Florida. The stratigraphy of these deposits has been constrained very precisely using seismic stratigraphy, biostratigraphy, sedimentology and sequence stratigraphy. The deposits are early to mid-Miocene in age. Three important phosphate-rich sedimentary sequences may be distinguished (FPS, OBS, BBS; Fig. 5.5). These are further divisable into 18 phosphate-rich horizons that reflect sea-level fluctuations of the fourth order. Each sequence lasted between 100,000 and 1,000,000 years.

Various phosphate components (authigenic phosphatized brachiopods, fish bones and teeth) were isotopically studied from sediments of the same age and from various bore holes. The Sr based, model ages showed variations of up to 2 million years for the same stratigraphic level with the amount of variation depending on the type of material used for analysis. The most homogeneous Sr ages were obtained from the phosphatic peloids; their ages differed by no more than 200,000 years. Investigations of McArthur et al. (1990) showed that this type of phosphate forms in isotopic equilibrium with seawater. However, this may also be the case for some of the other phosphatic components investigated. As these other components can also be of detrital origin, their model ages are less relevant then those of the phosphatic peloids which formed "in-situ".

Sr and Nd isotopic compositions determined on phosphatic peloids of the 'FPS-1' horizons are displayed in Fig. 5.5. The $^{87}Sr/^{86}Sr$ ratios decrease rather constantly with increasing depth and age and so reflect the seawater $^{87}Sr/^{86}Sr$ curve for this time interval. Comparing these ratios with the seawater $^{87}Sr/^{86}Sr$ evolution curve for this time yields an age of formation of 18 to 19 Ma for these peloids which is precisely the same age that can be ascertained using biostratigraphic methods.

Samples from the uppermost part of the profile, which has been interpreted by sedimentologists as a Pleistocene-Holocene condensation horizon, show a clearly homogeneous Sr isotope composition providing an age of about 17.2 Ma. This indicates that the phosphate here has been reworked and originated in the Miocene beds. The isotope investigations of the whole profile make it possible for us to make some important sedimentologic and stratigraphic conclusions.

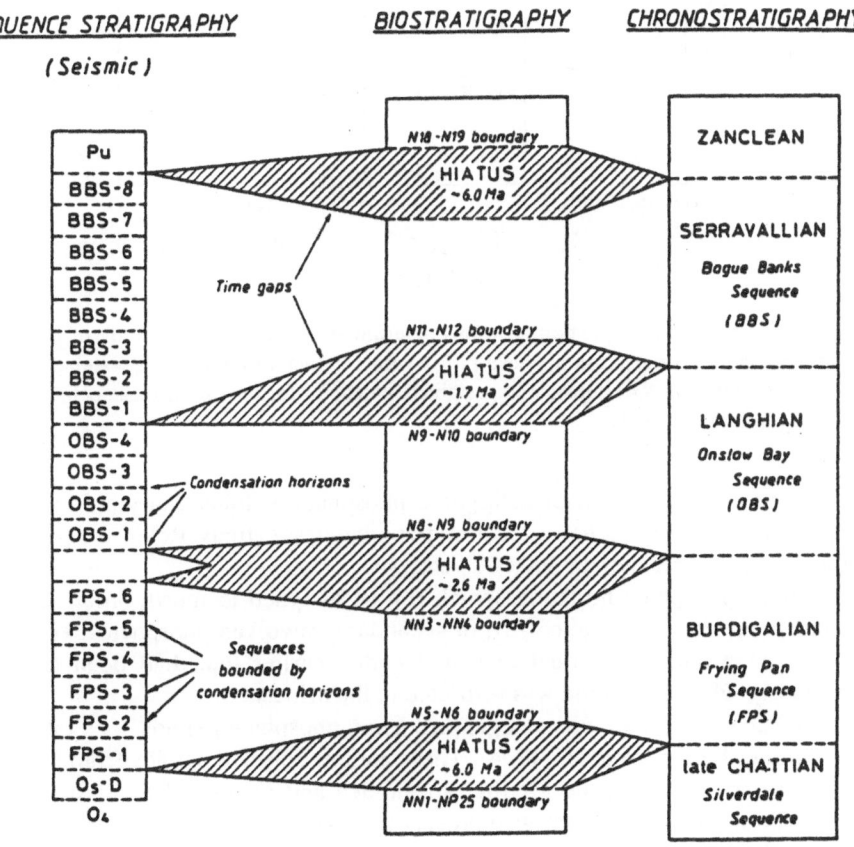

Fig. 5.5. Sequence stratigraphy of the largest phosphate deposits worldwide: North Carolina: the result of detailed seismic, biostratigraphic and sedimentologic investigations. Three important phosphate-rich sedimentary successions can be seen (FPS, OBS, BBS). These can be further subdivided into 18 smaller phosphate bearing sequences reflecting sea-level fluctuations of the fourth order. (after Riggs 1984)

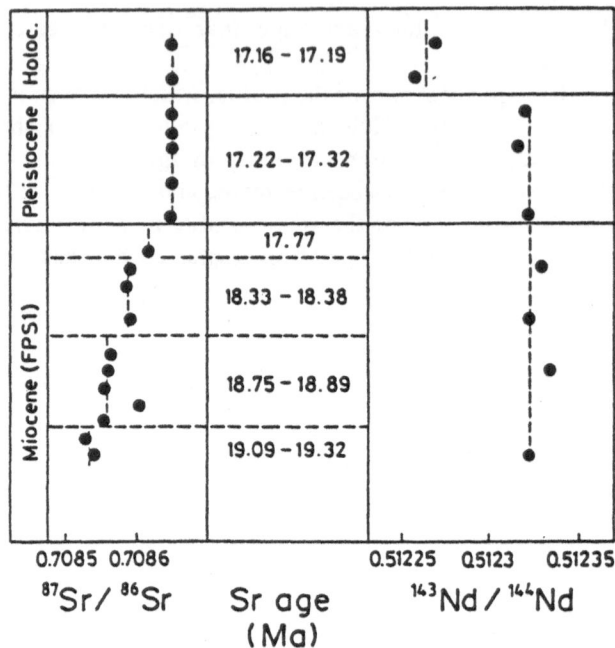

Fig. 5.6. Sr and Nd isotopic compositions determined on phosphatic peloids of the FPS-1 level (Fig. 5.5). Comparison of these ratios with the secular trend of seawater Sr isotope evolution (Figs. 5.3, 5.4) allows us to establish an age of formation for these peloids of 18-19 Ma. (Stille et al. 1994)

1. The Sr isotope system of authigenic phosphatic peloids can yield extremely precise age information for microstratigraphic investigations. Relative ages can be attained with a precision of 300,000 years.
2. Various phosphate horizons, which can be assigned to fourth order sea-level fluctuations were found to consist of secondary, reworked phosphates. However, phosphate production both began and ended earlier than had been assumed previously. Phosphatization was restricted to the Miocene.

The capacity to date phosphogenesis using phosphate peloids was taken one step further by Jacobs et al. (1994). These authors selected peloids from Miocene sediments of Malta in the Mediterranean Sea and compared their Sr isotopic compositions with data from well preserved foraminifera from the same sections and the data from the Miocene phosphate deposits on the other side of the Atlantic already mentioned. In this study, all peloids could be allotted Sr isotopic ages that were consistent with the foraminifera Sr isotopic ages and the microfossil-based biostratigraphic ages, showing that reworking was insignificant. Phosphogenesis could be dated at between 24 and 16 Ma, i.e. far longer than in North Carolina. Because of the good age constraints of these Maltese phosphates, work was also begun on their Nd isotopic composition (see Sect. 5.2.4) to shed light on the role of palaeo-oceanography on phosphogenesis.

5.1.2 The Role of Diagenesis in Sr Isotope Stratigraphy

Some mention has already been made of the problems encountered in Sr isotope studies due to diagenetic alteration. As one moves back through geologic time, analyzing ever more ancient and possibly more altered rocks, it becomes necessary to constrain or quantify this diagenetic disturbance. Such an approach has proved successful during several recent large-scale studies, most notably those of the Bochum/Ottowa research groups. Concentrating on what are clearly the most likely fossils to have retained seawater isotopic compositions, i.e. microscopically pristine, low-Mg calcite belemnites and brachiopods (other workers have concentrated on foraminifera for the Cenozoic), they have embarked upon an immense exercise to constrain precise trends in seawater Sr, C and O isotope ratios through the Phanerozoic era (<545 Ma), by systematically removing the ugly spectre of diagenetic alteration. Few studies have enlightened us more on how to constrain seawater Sr isotope ratio in past oceans than the early Sr isotope studies of Precambrian rocks, pioneered by Veizer (e.g. Veizer, 1983) and so we will concentrate on this period and the reconstruction of this part of the 'Sr isotope curve' in the following section.

The Sr isotope composition of seawater through Precambrian time (>545 Ma) is much less well known than for the Phanerozoic era. This has several reasons. Precambrian rocks are less common, more likely to be metamorphosed or to be heavily recrystallized (the result of often deep burial) or are dolomitic, in which case the carbonate components are likely to have lost more than 90% of their original strontium Veizer (1983). Our knowledge of Precambrian stratigraphy is also much less sure and the estimated ages of sedimentary suites liable to large errors. Correlation is hampered to such an extent that Sr isotope stratigraphy coupled with C isotope stratigraphy is often called upon to arbitrate in stratigraphic disputes. The absence of shelly fossils in the Precambrian, whose extent of diagenetic alteration can sometimes be assessed more easily (both petrographically and chemically), forces us to analyze bulk carbonate sediments. Constraining the original seawater Sr isotope ratio using bulk carbonate-rich rocks is an unfortunately tricky and time consuming option requiring the complimentary application of a battery of petrographic and geochemical approaches.

If we take another look at Burke et al.'s curve (Fig. 5.2), we can see that there is a great deal of scatter in the data although the Sr isotopic ratio of seawater ought to have been the same in all the oceans at any one time due to the long residence time of Sr and by comparison with today. Veizer (1983) discusses the problem of selecting the most suitable carbonate material for reconstructing real trends in seawater Sr isotopic ratio. This author was able to find important parameters which permit the identification of 'most altered' samples in a sample set. He and others have demonstrated that samples dissolved in HCl for analysis displayed positive correlation between the amount of detritus and the $^{87}Sr/^{86}Sr$ ratio. This implies that radiogenic Sr can be leached from detritus by a strong acid such as HCl. This Sr most probably derives from the clay fraction which contains relatively high Rb/Sr ratios. The decay of ^{87}Rb to ^{87}Sr means that unintentional leaching of even authigenic clays wil tend to skew Sr isotopic ratios higher, away from seawater. Dissolution using a weaker acid, possibly buffered around a relatively gentle pH, may help solve this problem as we will see in examples to come. However, some Sr will always be leached from detritus or later diagenetic carbonate phases and so it is equally important to check the purity of a sample as well as the type of detritus present.

Veizer (1983) observed correlations between Mn, Sr and Ca concentrations, and ^{87}Sr/^{86}Sr ratios. This relationship was displayed in a three dimensional diagram (Fig. 5.7). Higher Sr isotope ratios, i.e. those, which most deviated from seawater, were found in samples with higher Mn and lower Sr contents. The authors explained this using distribution coefficients. They argued that dissolution and recrystallization during diagenesis or low grade metamorphism would alter elemental abundances and isotopic ratios in a predictable fashion.

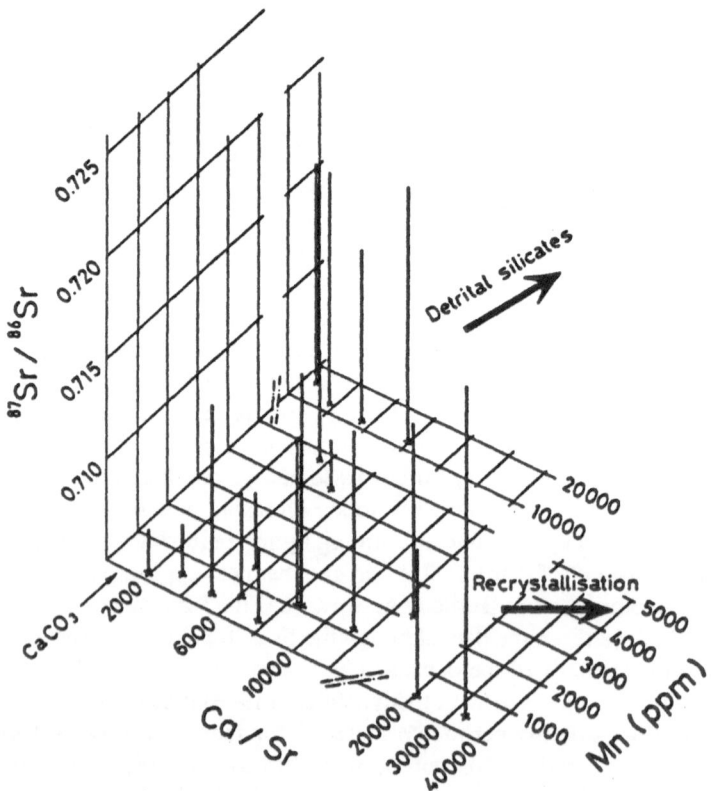

Fig. 5.7. Three dimensional ^{87}Sr/^{86}Sr - Mn - Ca/Sr diagram for the identification of uncontaminated marine carbonate (Veizer et al. 1983). Only marine carbonates situated in the lower left hand corner with low Mn contents, and Ca/Sr ratios yield a Sr isotopic ratio identical to seawater.

Sr has a lower distribution coefficient in calcite than Mn (<1 and >1, respectively) and so Sr tends to be lost during diagenesis while Mn tends to be incorpororated into the precipitating calcite depending on availability. On deposition, most carbonates have minimal Mn contents and contain up to 3-6000 ppm Sr. Veizer et al. calculated that recrystallization of a calcite in an open diagenetic system would result in Sr contents of around 400 ppm today. Such Sr loss may be accompanied by strontium isotopic exchange. Using the same principle, Mg tends to be lost during dissolution and reprecipitation events, while Fe is gained. Of course, loss of Sr from the system does not necessarily have to alter the isotopic ratio of the recrystallized carbonate. The effect of such trace element exchange depends on the closedness of the system and the presence of other sources having possibly different Sr isotopic ratios than seawater (hydrothermal fluids, detritus, Sr from older carbonates undergoing dissolution). Such diagenetic indicators may allow us to constrain initial Sr isotopic ratio (i.e. seawater) by constructing diagenetic trends.

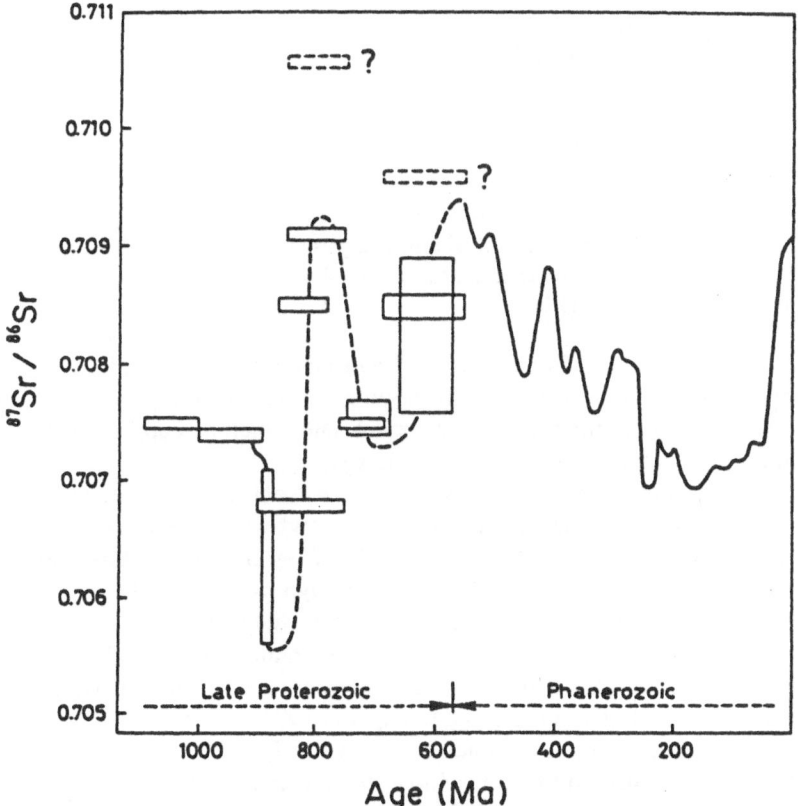

Fig. 5.8. The Sr isotopic composition of late Precambrian seawater. The lateral error bars represent stratigraphic uncertainties. The spread in the data is caused by reasons discussed in the text. (after Veizer and Compston 1974, 1976; Burke et al. 1982)

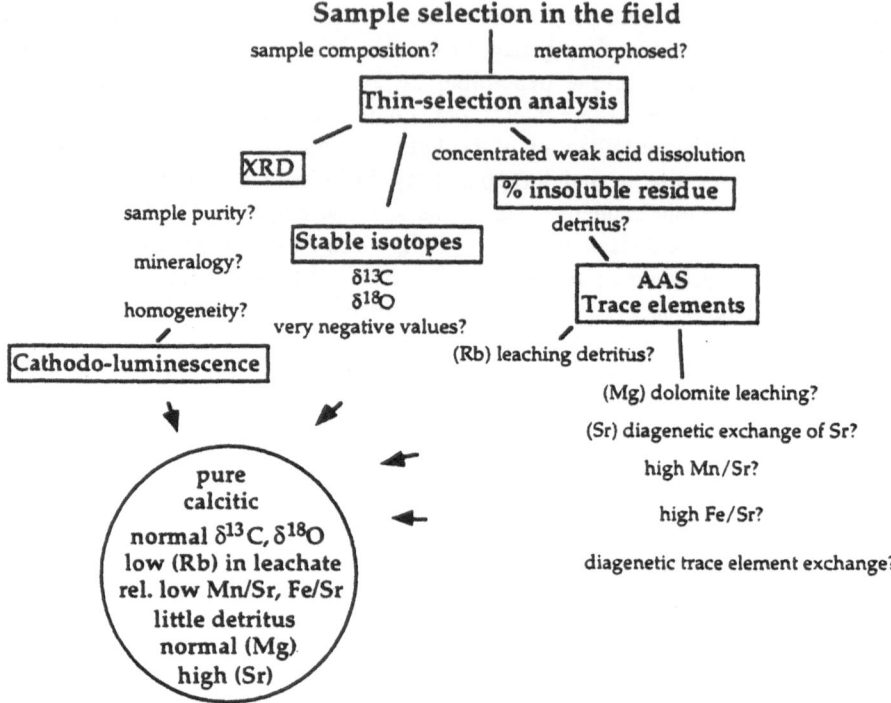

Fig. 5.9. Initial sample screening for Sr isotope analysis. This work-plan for the analysis of bulk carbonate rocks reflects the experience of several authors. The ringed part represents the ideal sample set. (Shields 1996)

Such trends (δ^{13}C, δ^{18}O, Mn/Sr, Fe/Sr, Sr/Ca or amount of detritus plotted against Sr isotopic ratio) have been used with varying success for samples, which have undergone alteration in the same environment, and which had approximately the same initial isotopic ratio, i.e. are of the same age.

A significant and pioneering study was carried out by Veizer et al. (1983). They analyzed late Proterozoic carbonate rocks and were able to constrain low values for seawater ^{87}Sr/^{86}Sr around 900 Ma. They interpreted this feature to imply a time of relatively high mantle input through hydrothermal cycling along the mid-oceanic ridge. It was important for this study, that samples could be compared from different parts of the World. The authors measured rocks from NW Africa and compared their data with literature data from various parts of Africa and from Australia (Veizer and Compston 1974; 1976; Burke et al. 1983). The Sr isotope ratios of these samples are shown in Fig. 5.8. Again, the data are marred by scatter caused by HCl leaching techniques that were too strong, high degrees of diagenetic alteration (e.g. high Mn/Sr ratios), open system diagenesis, presence of dolomite, and also the inevitable stratigraphic uncertainty. The low ^{87}Sr/^{86}Sr mantle 'event' which may or not be comparable in origin to the low ratios for the late Permian period and early Mesozoic era (Fig. 5.2.) was later confirmed by the work of Asmerom et al. (1991). This latter study is marked by samples with very high Sr contents and internally consistent stratigraphic trends in Sr isotope ratio and as such can be regarded as the beginning of our well constrained Precambrian Sr isotope record. Further additions to our knowledge of late Precambrian Sr isotopic evolution were made by Derry et al. (1992) and can be seen summarised in Brasier et al. (1996).

Subsequent studies have begun to select samples very carefully before measurement to avoid measuring altered samples (see Fig. 5.9), although it is often necessary to analyze some poorly preserved samples in order to establish diagenetic trends (Fig. 5.10). Each sample set is different and has reacted to possibly different non-seawater sources of Sr. Thus, it is essential to measure as many additional parameters as possible to get the most out of any data set.

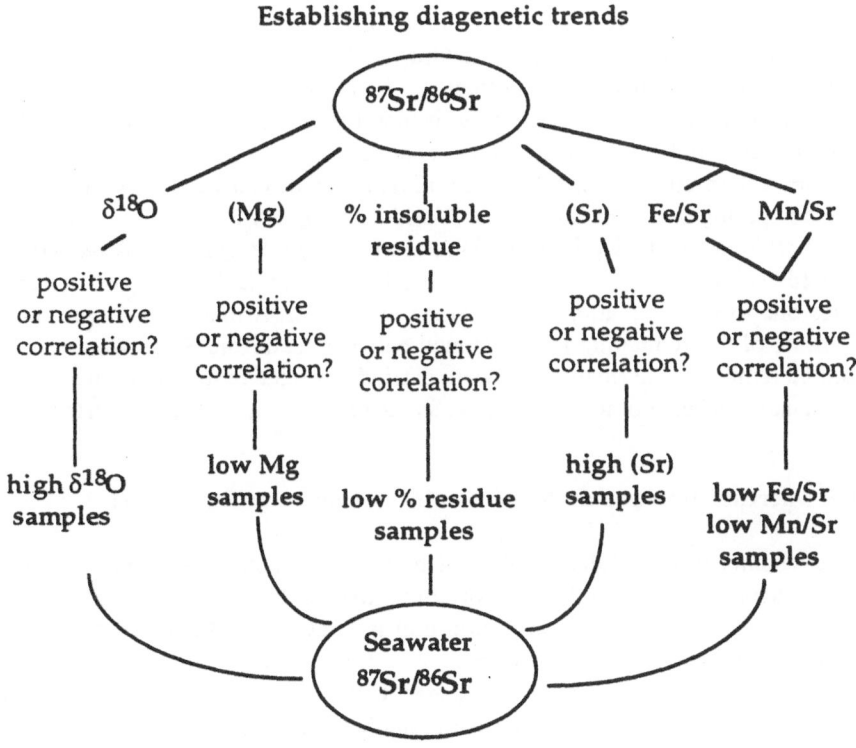

Fig. 5.10. A scheme for establishing how diagenetic trace element exchange has affected the Sr isotope composition of a group of samples from one stratigraphic level. As each parameter is more complicated than shown here, it is important to apply as many parameters as possible to understand the geochemical evolution of the rocks more fully. (Shields 1996)

As a result of several studies, which have used the above approaches, it has become clear that scatter in Sr isotope data may derive from a number of sources:

1) Leaching technique. To avoid measuring more radiogenic Sr which derives from clay minerals (authigenic or detrital) or Sr from detritus, it is necessary to apply the gentlest leaching technique possible for that particular mineralogy. For limestones, this is likely to involve an initial cleaning of easily leachable ions by ammonium acetate or very dilute HCl and thereafter dissolution in acetic acid buffered to a mild pH (about 4.5) by an acetate buffer. Dolomites and apatites may require an HCl leach and this is one further reason why they yield such variable results.

2) Rb correction. Rb which is leached from the carbonate or elsewhere will have undergone radioactive decay to ^{87}Sr since the time of formation. Provided this ^{87}Sr has not escaped from the system, it will tend to make the measured ^{87}Sr/^{86}Sr ratio too high and this effect must be corrected for. The older the samples, and the higher the initial Rb/Sr ratio, the greater is the effect! However, such a correction involves several assumptions and it is best to avoid samples requiring any extra correction of this nature.

3) Diagenetic exchange. This effect can be minimalized by the establishment of diagenetic trends (Fig. 5.10) and by the careful selection of samples (Fig. 5.9). These aspects will become clearer in the next section.

4) Errors in stratigraphic correlation. This effect can be highly significant when comparing data from around the World (see Fig. 5.8).

5) Different standard values. All laboratories calibrate against an international standard: NBS SRM 987. Sometimes batches of standard may give different values due to contamination and machine effects. McArthur (1994) recognized that some of the remaining inconsistency between Cenozoic Sr seawater curves could be removed if normalization were carried out relative to a modern seawater shell standard instead of to this artificial standard as a result of analytic artefacts.

5.1.3 Microstratigraphy at the Precambrian/Cambrian Boundary?

Improvements in our understanding of how the Sr isotopic system may react during diagenetic alteration brings us right up to date. It is now possible to use the Sr isotopic system as a high resolution tool in stratigraphy for much of the past 800 Ma. Its power lies in the assumption of isotopic homogeneity during times of good ocean circulation and the possibility of combination with other chemostratigraphic parameters such as the stable isotopes carbon and oxygen. One recent study used this flexibility to make stratigraphic correlation of the Precambrian / Cambrian boundary easier (Brasier et al. 1996). As this boundary marks the introduction of animal shells in the geologic record, it was necessary to measure bulk carbonate rock for all geochemical parameters.

This international group of authors, representing part of the IGCP (International Geological Correlation Programme) group working on the formal definition of the boundary, measured C and O isotope ratios, Fe, Mn, Mg, Rb and Sr contents of

HCl-soluble fractions and Sr isotope ratios of the acetic acid soluble fraction for over 100 bulk limestone rocks. Samples underwent an initial selection procedure similar to that described in Fig. 5.9. After isotopic analysis, covariation of diagenetic trends with Sr isotope ratios were assessed as in Fig. 5.10. Fig. 5.11. shows one such trend.

Sr contents proved to be the most reliable diagenetic alteration proxies in this case with other parameters generally showing no clear trends. Other studies have concluded that Mn/Sr ratios or $\delta^{13}O$ values were more reliable in their cases. This relationship is however unlikely to exist always as stable isotope exchange is not necessarily accompanied by Sr isotope exchange, and Mn concentrations in diagenetic fluids are largely controlled by other factors such as redox potential other than simply the openness of the system.

The predictable increase in $^{87}Sr/^{86}Sr$ ratio with decreasing Sr content (Fig. 5.11) allowed subtle trends in initial seawater $^{87}Sr/^{86}Sr$ to be established. Fig. 5.12 displays data from the upper (filled circles) and the lower parts (open circles) of one sedimentary section, the southern Bayan Gol section, W. Mongolia (see Brasier et al. 1996).

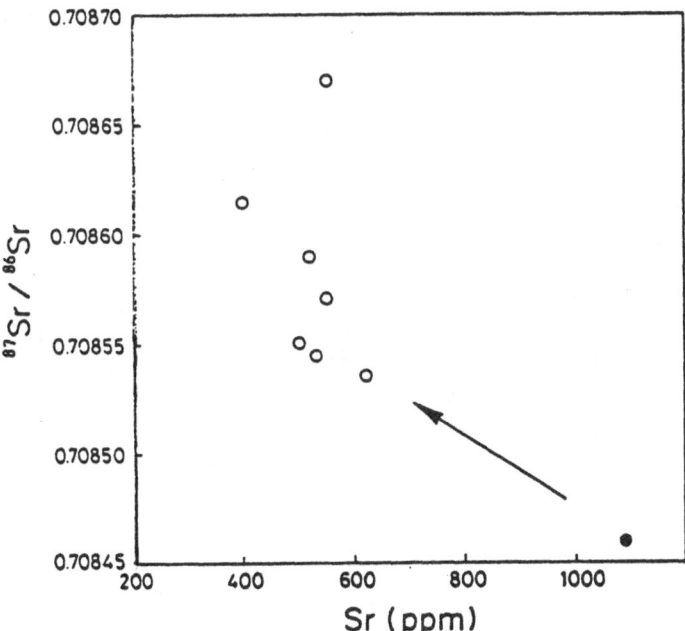

Fig. 5.11. $^{87}Sr/^{86}Sr$ ratios plotted against Sr content for one section close to the Precambrian / Cambrian boundary in W. Mongolia (Shields 1996). As Sr was being lost from the carbonate, exchange with more radiogenic diagenetic fluids caused the $^{87}Sr/^{86}Sr$ ratio of the crystallizing calcite to increase. Thus, lowest $^{87}Sr/^{86}Sr$ represent our closest approximation to seawater composition.

Both sets of data show clear diagenetic trends recording predictable incorporation of radiogenic Sr during diagenesis. Thus, the lowest ratios for each stratigraphic level are our best indicator for initial seawater $^{87}Sr/^{86}Sr$ ratio. The best preserved samples, i.e. those with the highest Sr contents, imply that seawater $^{87}Sr/^{86}Sr$ ratio decreased from <0.7085 to <0.7082 close to the Precambrian / Cambrian boundary.

After considering the data set as a whole, the authors decided to divide the sample set into two groups: 'least altered' or 'best preserved', and 'altered'. Altered samples were those whose Sr isotope ratios were considered less likely to represent seawater; they contained less than 600 ppm Sr and had large insoluble residues. As expected, least altered samples consistently had the lowest initial $^{87}Sr/^{86}Sr$ ratios as light Sr from volcanic material is a less common factor during diagenesis than isotopically heavier Sr from clay minerals. These best preserved samples defined a rise in seawater Sr isotope ratio from around 0.7067 to 0.7085 (around 700 Ma to 545 Ma) at the introduction of the very first shelly fossils (sponges and simple tubes) in western Mongolia.

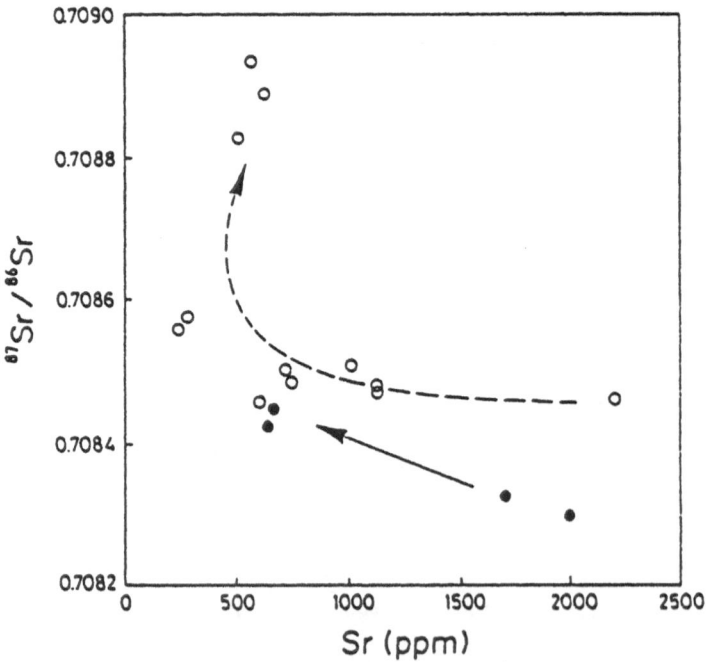

Fig. 5.12 $^{87}Sr/^{86}Sr$ ratios shown against Sr content for two stratigraphic levels of the Bayan Gol Precambrian / Cambrian section in W. Mongolia (Shields 1996). Samples from the two levels which have suffered similarly minor Sr loss yield different $^{87}Sr/^{86}Sr$ ratios, suggesting a real, but subtle change in seawater $^{87}Sr/^{86}Sr$ through time from 0.7085 to 0.7083.

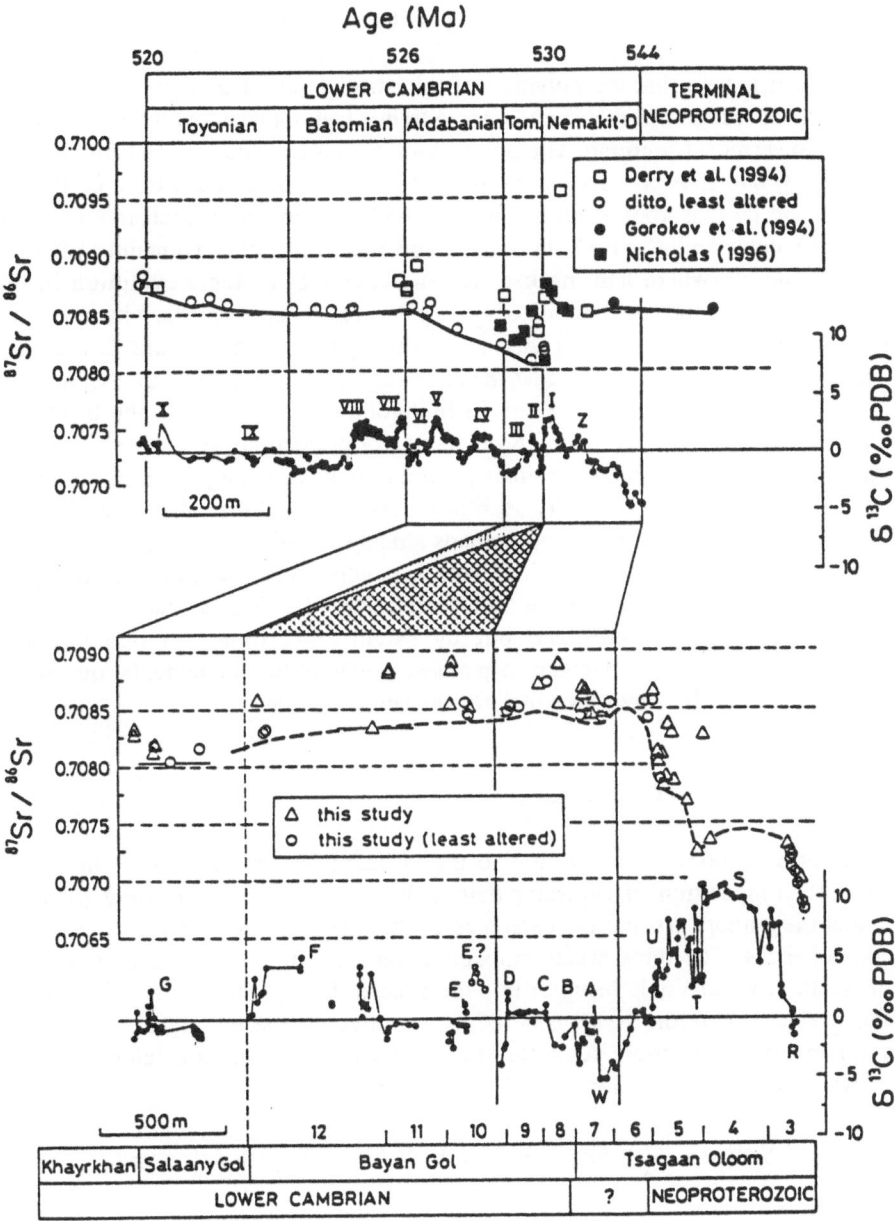

Fig. 5.13. Curves of published Sr and C isotopic data across the Precambrian / Cambrian boundary. Above: SE. Siberia from Brasier et al. (1994) and authors mentioned in figure; below: W. Mongolia from Brasier et al. (1996) and Shields (1996). Comparison of the two sets of curves reveals isotopic features in the Mongolian data set not apparent in Siberia. (Brasier et al., 1996) conclude that there may be an unconformity in the boundary sections of Siberia, which might mask the true, more gradual course of the Cambrian explosion.

Thereafter, seawater $^{87}Sr/^{86}Sr$ appears to have dropped gradually to around 0.7081 (Fig. 5.13). At the same time, C isotope ratios rose from around -4 to +5 ‰ PDB. It is here that the potential of this stratigraphic technique can be seen. The authors recognized that such isotopic trends had not been recorded elsewhere for the lowermost Cambrian. The other major database available was from Siberia (Fig. 5.13) and showed only an abrupt change in Sr isotope ratio from 0.7085 to 0.7080 and no C isotope values as high as +5‰. Although inconclusive on their own, the data would permit the conclusion that a significant unconformity in Siberia and elsewhere had masked the true course of biologic evolution in the Early Cambrian. Indeed sections of this age which do contain sharp increases in faunal diversity also show signs of hiatus within them and are commonly condensed. The authors argue that the relatively gradual nature of the 'Cambrian explosion' in shelly fossil diversity in Mongolia may be closer to the truth and that there was no massive faunal turnover at this point as suggested elsewhere. In this case, the fortunate tectonic situation of a rapidly subsiding basin in Mongolia avoided the major unconformity seen in most other carbonate shelf successions. This helped smooth out the isotopic trends already recorded elsewhere.

Sr isotope stratigraphy, combined with C isotope-, bio- and event stratigraphy can provide us with high resolution correlation of events as far back as the Neoproterozoic. Not only that but by establishing trends in geochemical parameters through time, we can help answer some of the fundamental questions regarding past global environmental change and evolution.

5.1.4 Mass Balance Calculations

Faure et al. (1965) were the first to try to quantify the fluctuations in the Sr isotopic composition of seawater that took place over geologic time. In their model the authors assume the existence of three sources of Sr, which are all the consequences of chemical weathering, either on the surface of the continents or on the seafloor which could bring Sr into the oceans. The resulting $^{87}Sr/^{86}Sr$ ratio of seawater can therefore be considered as a product of the mixing of these three components. The authors suggested the following mass balance calculation for seawater isotopic composition:

$$\left(^{87}Sr/^{86}Sr\right)sw = S\left(^{87}Sr/^{86}Sr\right)s + V\left(^{87}Sr/^{86}Sr\right)v + M\left(^{87}Sr/^{86}Sr\right)m$$

S,V, and M represent that proportion of Sr which derives from crustal (s), volcanic (v) and marine carbonate (m) sources respectively (s+v+m = 1).

$(^{87}Sr/^{86}Sr)$ sw   : seawater isotopic ratio
$(^{87}Sr/^{86}Sr)$ s   : Sr isotopic ratio from volcanic rocks (0.704)
$(^{87}Sr/^{86}Sr)$ v   : Sr isotopic ratio of crustal rocks (0.720)
$(^{87}Sr/^{86}Sr)$ m   : Sr isotopic ratio of marine carbonate and sulfate

It is interesting that the authors did not consider an oceanic hydrothermal component in their mass balance, although this can be taken as part of "V". The diagram in Fig. 5.14 displays seawater Sr isotope ratio plotted against input of volcanic strontium. This illustrates the above mass balance equation. Values representing the proportion of Sr from marine carbonates (m) and crustal rocks (s) are similarly displayed. Parallels represent fractions of input from s and m. The equation defines a triangular field on which can be found the possible $^{87}Sr/^{86}Sr$ ratios of seawater, bordered by the cases where one of the sources inputs nothing into the oceans, i.e. where V, S or M = 0.

Seawater $^{87}Sr/^{86}Sr$ ratio varied between 0.7065 and 0.7092 during the Phanerozoic. These boundary values are displayed: Point A represents the very improbable case whereby no Sr derives from the dissolution of marine carbonate (M = 0). In this case, V = 0.7 and S= 0.3. In other words, 70% of the strontium derives from volcanic sources and 30% from crustal sources. As M must be greater than 0, it can be assumed that these relative proportions represent maximum values and are really lower.

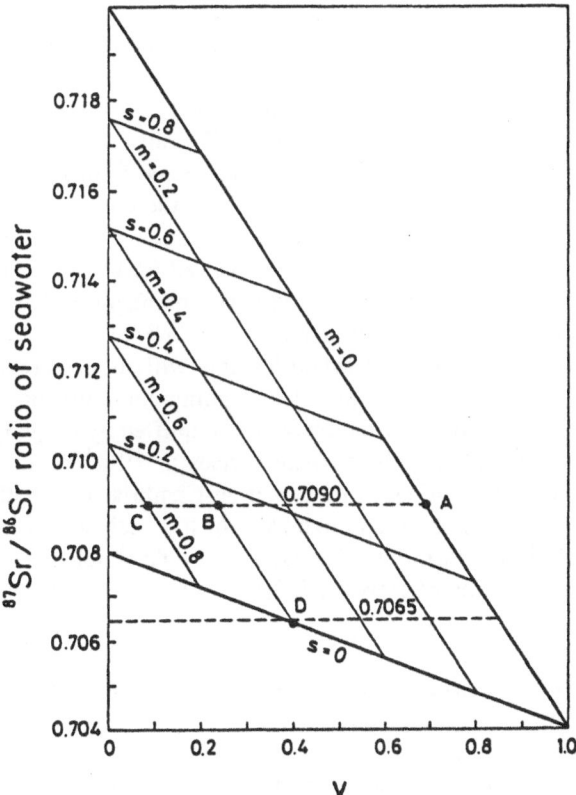

Fig. 5.14. Seawater Sr seen as a mixture of crustal (S), volcanic (V) and marine carbonate (M) sources. (Faure et al. 1965)

A more probable situation is represented by point B. In this case M = 0.6, V = 0.25 and S = 0.15. The very high Sr contents of marine carbonates (up to 6000 ppm), their high solubility and their worldwide distribution on the continents would tend to suggest that a significant proportion of the strontium in the ocean comes from this source. Apart from weathering, strontium is also released from carbonates in another way. Elderfield and Greaves (1981) observed that submarine carbonate recrystallization can exert a strong influence on the seawater isotopic composition. Sr is released during the diagenetic recrystallization of aragonite to calcite. Our estimation that up to 60% of the strontium that is found dissolved in the oceans originates from marine carbonate does not appear improbable. This would mean that 25% of the strontium derives from volcanic, basaltic sources and only 15% from crustal, sialic rocks at the present day.

In the Permian and Jurassic periods and in the late Precambrian the Sr seawater curve reached minimum values (Figs. 5.2; 5.8). If we assume that the proportion of Sr deriving from marine carbonate was 60%, we can calculate that at these times, the proportion of strontium from the mantle was 40% and that from crustal, sialic sources approached zero (point D). This mantle influence could have been provided by enhanced rates of ocean spreading and intensive exchange between seawater, newly produced oceanic basalts and hydrothermal fluid phases. It is possible that this enhanced activity had some connection to the breaking up of the supercontinents Pangaea and Rodinia (Proto-Pangaea) respectively, which are likely to have occurred at this time (Piper, 1982). We must remember that such an hydrothermal event can only be recognizable using $^{87}Sr/^{86}Sr$ if there was relatively insignificant input of radiogenic Sr from the continent, i.e. relatively low weathering rates. Low weathering rates were made possible by the existence of a peneplained supercontinent as has been envisaged for both late Permian times (Pangaea) and Late Precambrian times (Rodinia). The beginning of the breaking up of a supercontinent may lead initially to low $^{87}Sr/^{86}Sr$ ratios because of the enhanced rifting associated with break-up but this will soon be reversed by the start of mountain building due to subsequent doming and collision. The erosion and weathering of granitic rocks, which have been uplifted by orogeny, will serve to increase seawater $^{87}Sr/^{86}Sr$. The two steep rises in $^{87}Sr/^{86}Sr$ in the geologic record (Cenozoic till today and latest Precambrian-Cambrian) are both considered to represent orogeny and erosion and exposure of the ancient granitic roots of a mountain belt (high $^{87}Sr/^{86}Sr$ ratios). The Himalayan-Tibet uplift. is considered to have controlled the evolution of seawater $^{87}Sr/^{86}Sr$ over the last 30 Ma.

5.1.5 The Influence of Groundwater on the Sr Isotopic Composition of Seawater

In order to understand long-term seawater isotopic variations better, Chauduri and Clauer (1986) compared these with the calculated and derived sea level fluctuations for the Phanerozoic of Vail et al. (1977).

Vail et al. (1977) were able to demonstrate that sea level fluctuations are correlatable with spreading rates at the mid-ocean ridge. Increased rates of spreading correspond to periods of transgression, while decreased rates of spreading correspond to times of regression. As high spreading rates are likely to be accompanied by enhanced hydrothermal activity, Chaudhuri and Clauer (1986) assumed that the $^{87}Sr/^{86}Sr$ ratio of seawater is controlled by the relative importance of transgression and regression. During times of high sea level for example, seawater $^{87}Sr/^{86}Sr$ ratio should be low.

However, sea level fluctuations and seawater $^{87}Sr/^{86}Sr$ ratios appear to be only partly correlated with each other (Fig. 5.15). In this respect, the lowering of the $^{87}Sr/^{86}Sr$ ratio and the beginning of transgression around 180-200 million years ago may conceivably be connected with the onset of crustal rifting in Pangaea as discussed in the last section (see Schaltegger et al., 1994; Fiechtner et al., 1992). From this one can assume that hydrothermal activity at this time was not only enhanced on the continent but also in the Panthalassa Ocean. The lowest $^{87}Sr/^{86}Sr$ ratios were reached during steadily rising sea level just before 120 Ma ago. The timing of this low point matches particularly well to the time of spreading, especially that associated with the formation of the first oceanic crust in the Tethys and the Atlantic. Using this model, however, we cannot explain very well the decoupled behavior of sea level fluctuations and $^{87}Sr/^{86}Sr$ ratios in the Cretaceous period. Here $^{87}Sr/^{86}Sr$ ratios increase as the transgression continued. In Permian-Triassic time, sea level fell as seawater $^{87}Sr/^{86}Sr$ reached a minimum. Both instances can be explained by the compensating input of strontium from the predominantly "sialic" continents. It can be assumed that there was relatively minor continental input of strontium during Permian / Triassic time. Lack of continental weathering would be enough to cause such a low $^{87}Sr/^{86}Sr$ ratio without the necessity for any marked increase in hydrothermal input. Similar arguments can be brought to bare to explain the anomaly in the Cretaceous, with the onset of rifting leading eventually to further mountain building and erosion, however such a solution is not entirely satisfactory.

Chauduri and Clauer propose another solution. They postulate the presence of an additional continental source, which would supply radiogenic strontium to the oceans. As we shall see, groundwater provides a finishing touch to explain some of the variations in Sr isotope composition during the Phanerozoic.

On the basis of an unfortunately limited number of measurements and calculations, it can be assumed that the volume of groundwater, which finds its way into the oceans makes up about $1.7 * 10^3$ km^3. $2 * 10^{12}$g of strontium are transported into the oceans by this route every year. If we assume steady state conditions in seawater, then the following mass balance may be applied:

$$Sr_{rw}R_{rw} + Sr_{gw}R_{gw} + Sr_{dc}R_{dc} + Sr_{ob}R_{ob}$$
$$= (Sr_{rw} + Sr_{gw} + Sr_{dc} + Sr_{ob})R_{sw}$$

Sr_{rw} = Contribution of Sr by river transport (2.21×10^{12}ga^{-1})

R_{rw} = Sr isotopic ratio of river water (0.7111)
Sr_{gw} = Sr groundwater flux (1.9 x 10^{12}ga^{-1})
R_{gw} = Sr isotopic ratio of groundwater (0.711)
Sr_{dc} = Sr flux from the recrystallization of carbonates (0.48 x 10^{12}ga^{-1})
R_{dc} = Sr isotopic ratio of the recrystallization flux (0.7084)
Sr_{ob} = Sr flux from oceanic basalts (1.26 x 10^{12}ga^{-1})
R_{ob} = Sr isotopic ratio of basalts (0.703)

The mass balance is similar to that of Faure et al. (1965). New additions are the contribution from groundwater Sr, Sr which is released by the diagenetic recrystallization of marine carbonate sediments, and Sr from submarine hydrothermal exchange.

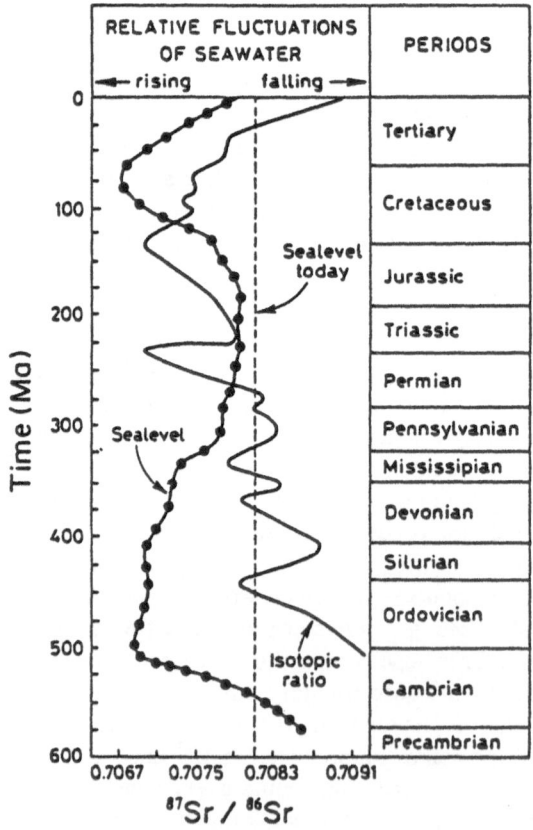

Fig. 5.15. Relationship between Sr isotopic variation in the oceans and sea level fluctuations through geologic time. Chauduri and Clauer (1986) were able to show that sea-level fluctuations and related hydrothermal seawater circulation are not always the majpr influences on the Sr isotopic composition of seawater. The authors postulate an additional source of Sr that supplies the oceans with relatively more radiogenic Sr: groundwater.

River waters, according to this newer equation, deliver strontium which is derived from continental, sialic rocks as well as marine carbonates and young volcanic rocks exposed on the surface of the Earth. Faure, in his equation, applied three different fluxes for these components. The groundwater flux calculations show that the global contribution of Sr into the oceans from groundwater would match that from rivers.

What kind of influence could groundwater have had during the Phanerozoic? An important factor that controls the amount of groundwater flow into the oceans is the variation in length of coastline and fluctuations in sea level. Fragmentation and the splitting up of continents or regression result in an increase in the length of coastline, whereas continent-continent collision and continent-island arc collision or transgression lead to a reduction in the amount of coastline. Transgressions or regressions would be likely to have the same effect on the Sr contribution from both river water and groundwater. Continental fragmentation or collision would however be likely to affect the Sr contribution from river water far less than that from groundwater.

Such mass balance considerations allow us to explain some of the more noticeable Sr variations in the Phanerozoic. For example, the rise in seawater $^{87}Sr/^{86}Sr$ ratio after the Cretaceous parallels the further breaking apart of the supercontinent Pangaea as well as eustatic regression. These mechanisms have the consequence that the flux of groundwater to the oceans increases in particular and with it, the flux of radiogenic, continental Sr. The Sr isotopic composition may have increased additionally due to the relative decrease in mid-ocean ridge hydrothermal activity which may have led to the regression.

The rise in $^{87}Sr/^{86}Sr$ ratio during the Cretaceous occurred during a period of sustained transgression and likely enhanced sub-aquatic hydrothermal activity, which would tend to lower the Sr isotopic ratio of seawater. Therefore, this rise is likely to relate wholly to a compensating increase in continental input of strontium. In Cretaceous times, 35% of today's land mass was covered by water and the supercontinent Pangaea was still more or less intact. The amount of coastline must have been a great deal shorter than that of today. As a consequence, much less continental strontium could have made it into the oceans than during the Cenozoic, although still much more than at the Jurassic-Cretaceous transition. In Jurassic-Cretaceous times, fragmentation of the supercontinent was limited to the barest minimum and so the contribution of groundwater strontium must have also been less. The $^{87}Sr/^{86}Sr$ ratios reached a minimum during this time.

Apart from a few fluctuations in seawater $^{87}Sr/^{86}Sr$ ratio that can be correlated with important orogenic phases, the Paleozoic is marked by a general fall from high to low ratios from the Cambrian period until the Permian (545 - 250 Ma). The formation of the supercontinent Pangaea led to a decrease in the groundwater flow into the oceans which may have caused a general decrease in the contribution of continental strontium.

There are few data for the Precambrian. What can be drawn from what little we know is that the breaking up of the >800 Ma old supercontinent (Rodinia) was also accompanied by a general rise in $^{87}Sr/^{86}Sr$ ratio, mirroring the post-Triassic story in the Phanerozoic. This rise was between 0.7057 and 0.7091, and was completed by the Middle Cambrian.

This discussion shows that $^{87}Sr/^{86}Sr$ minima are not always necessarily the direct result of enhanced hydrothermal activity. Very low $^{87}Sr/^{86}Sr$ ratios may also be the result of a low groundwater, and/or river flux, caused possibly by the formation of a supercontinent.

5.2 Nd Isotopic Composition of Seawater: Tracer of Water Circulation Patterns

Neodymium (Nd) is also present as a trace element in seawater. However, Nd concentrations are far lower than for Sr varying between $0.41 * 10^{-6}$ and $4.21 * 10^{-6}$ mg/L. The mean Nd concentration lies around $2.58 +/- 0.23 * 10^{-6}$ mg/L making it 3 million times less abundant in the oceans than Sr. Despite, the low concentrations direct isotopic measurement of Nd in seawater has proved possible. As a result of these analyses we now know that the $^{143}Nd/^{144}Nd$ isotopic ratio in seawater is not homogeneous as is the case with Sr. The Nd isotopic ratio varies not only from ocean to ocean but also within a single ocean basin (Fig. 5.16).

On the basis of investigations by Piepgras et al. (1979) and Piepgras and Wasserburg (1980, 1982) εNd values (for definition see Sect. 1.2) of -12 +/- 2 could be determined for the present day Atlantic Ocean, -3 +/- 2 for the Pacific Ocean and -8 +/- 2 for the Indian Ocean. The different Nd isotopic compositions between oceans are the result of the isotopically different sources of Nd for these oceans and the short residence time of Nd in seawater, which is far shorter than that for Sr and is also shorter than the time of around 3000 years required for the oceans to become thoroughly mixed (see also Sect. 5.2.1).

The relatively high εNd value for the Pacific ocean can be attributed to the juvenile crust, which surrounds the Pacific and also the high production rate of volcanic rocks within the Pacific itself. The relatively low values for the Atlantic result from the existence of old continental blocks (cratons) on either side of that ocean (Brazil, Canada, Fennoscandia, Greenland, West Africa); the crust is marked by low Sm/Nd ratios, Sm being preferentially incorporated into the mantle. The εNd values show that more than 90% of the Nd in the Atlantic is of continental origin, while this source represents only 60% of the whole for the Pacific. As the Nd concentrations in seawater are far lower than those for Sr, the direct determination of seawater Nd isotope composition is extremely difficult. In order to get around this problem, it is necessary to analyze rocks and mineral components enriched in Nd that are likely to have incorporated and preserved a seawater isotopic signature.

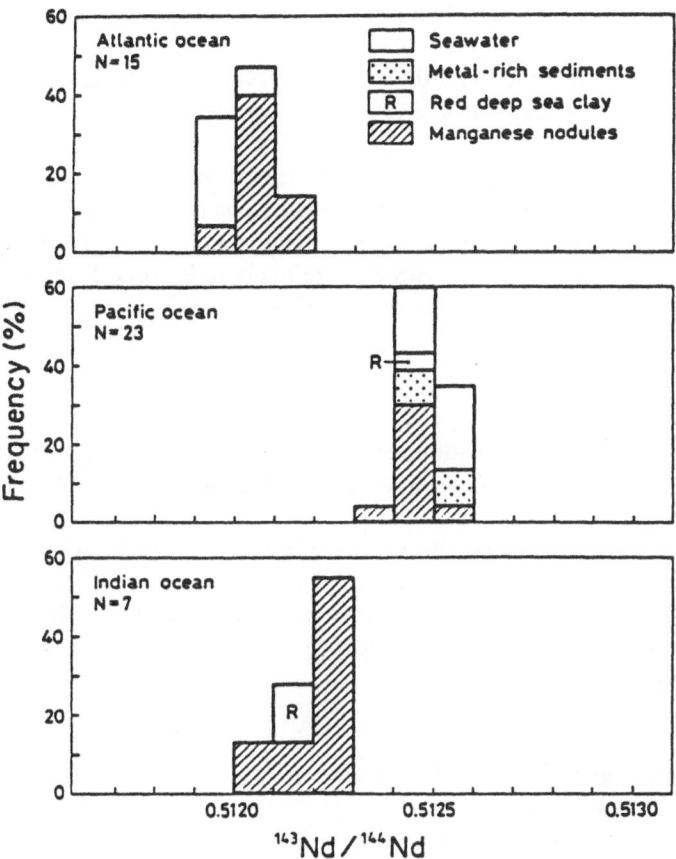

Fig. 5.16. Present day Nd isotopic composition in the oceans.

The first study that looked at the Nd isotopic composition of the Pacific was carried out by DePaolo and Wasserburg (1977) on phosphate fish bone splinters. These yielded a εNd value of -9.2. Today we know however that the actual value should lie between 0 and -5 and is therefore far higher than that determined by DePaolo and Wasserburg. It is possible that these authors analyzed continental detritus along with fish phosphate or that diagenetic processes had altered the isotopic composition of the bone by exchanging with detritus. If this pioneering study did indeed measure unaltered bone phosphate, then a further interesting possibility presents itself: perhaps the fish grew up in the Atlantic ocean incorporating therefore an Atlantic isotopic signature but had died in the Pacific.

O'Nions et al. (1978) were able to determine relatively homogeneous Nd isotopic ratios from metal-rich sediments and ferromanganeses nodules from the floor of the Pacific ocean (εNd: -4.3 to - 1.6). Piepgras et al. (1979) then went on to investigate not only oceanic ferromanganese nodules but also managed to carry out direct Nd isotope analyses of seawater for the first time. They were able to show that manganese nodules do indeed incorporate the Nd isotopic composition of the surrounding seawater and so could be used to establish temporal changes in seawater εNd.

Studies on seawater today as well as studies on marine authigenic minerals from ancient oceans have shown that these oceans also were never as well mixed with respect to Nd as to Sr. As kinetic fractionation cannot be the reason, their respective residence times in seawater are likely to be very different. Let us look into the calculation of the residence time of Nd.

5.2.1 Calculation of the Residence Time of Nd in Seawater and of the Time Necessary for the Thorough Mixing of Two Oceans or Water Bodies

The observable differences in Nd isotopic composition between ocean bodies allows us to calculate mass balances for these oceans as well as to infer the time necessary for the water bodies of two oceans to become thoroughly mixed. Let us begin with the derivation of the residence time of Nd in seawater. The residence time (τ) is defined as follows:

residence time (τ) = size of the reservoir(s) / flux
where,
flux = input of a certain mass / time

This equation allows us to calculate for example the residence time of water in the World's oceans. The oceans contain some 1.4×10^{21} kg of seawater. The flux from the rivers consists of about 4.2×10^{16} kg/year. The residence time can therefore be estimated at about 3×10^4 years. If the oceans are well mixed, i.e. if there is vigorous circulation controlled by high temperature gradients from the poles to the equatorial regions, the water from rivers and their dissolved loads can be spread throughout the World's oceans. The time for complete mixing of the oceans is estimated at about $1.5 - 3 \times 10^3$ years (ocean mixing time). From this we can see that if an element's residence time is shorter than 3000 years, significant variations will be established in elemental concentrations and isotopic ratios depending on the proximity to source and the isotopic inhomogeneity between the various sources. Let us concentrate now on the residence time of Nd in seawater.

As we have seen, the Nd budget is largely controlled by the input of Nd from the continents. The average Nd concentration in the 'effectively dissolved' load is about 16.4 ppt, whereas that in the ocean is only about 3.5 ppt (Table 4). The residence time for Nd may be calculated as shown in equation I:

$$t(Nd) = \left(\frac{WS * CS}{FR * CR}\right) = \left(\frac{WS * CS}{F(Nd)}\right) = \left(\frac{1.4 * 10^{21}[kg]\ \ 3.5 * 10^{-12}[kg/kg]}{4.2 * 10^{16}[kg/yr]\ \ 16.4 * 10^{-12}[kg/kg]}\right)$$

= c. 7100 years

where   WS   : mass of water in the oceans
              CS   : concentration of an element in the ocean
              FR   : flux of water from the continents
              CR   : concentration of an element in this water
              F(Nd) : flux of Nd from the continent.

Goldstein und Jacobsen's (1988) estimated Nd residence time of about 7100 years is high in comparison with other calculations and does not seem to correspond with what we already know about the isotopic inhomogeneity of Nd in the oceans; 7100 yr is two to four times the length of the estimated ocean mixing time. It is possible that the concentration of 'effectively dissolved" load (16.4 ppt) was highly underestimated (Goldstein and Jacobsen, 1988; Sect. 3.2). Goldstein und O'Nions (1981) managed to calculate a residence time of 3100 yr and Piepgras et al. (1979) derived a Nd residence time of about 400-2000 yr by applying the Na residence time and the Na/Nd ratios in river waters and seawater:

$$\tau(Nd) = \tau(Na)\left[\frac{(Na/Nd)R}{(Na/Nd)S}\right]$$

where R stands for river water and S for seawater

Thus, the residence time for Nd is anyway much shorter than for Sr ($\approx 3 - 5 \times 10^6$ years). Let us try to derive the time necessary for two oceans (e.g. the Pacific and the Atlantic) to become thoroughly mixed. We need to start from the mass balance equations for both oceans and assume steady state conditions. As the seawater Nd isotopic ratio strongly reflects input from the continent let us assume for simplicity's sake that the input from the mantle is negligibly small. The following factors are important for the construction of the necessary equation system (after Piepgras and Wasserburg 1980):

1) Input into ocean 1 (flux 1): FA1;   kg Nd/year
     Input into ocean 2 (flux 2): FA2;   kg Nd/year

2) Sedimentation rate in Ocean 1: F_{S1};   kg Nd/year
     Sedimentation rate in Ocean 2: F_{S2};   kg Nd/year

3) Rate of exchange between ocean 1 and ocean 2: $W_{1,2}$

$$W1,2 = \left(\frac{W1}{T}\right)$$ W1: mass of water or volume; T : rate of exchange

$C_1 \cdot W_{1,2} = [\text{kg Nd/year}]$

For "steady state" conditions (amount of water input = amount of water output) the following are valid:

$$F_{A1} + C_2 W_{1,2} = F_{S1} + C_1 W_{1,2} \quad \text{(II)} \quad \text{(in ocean 1)}$$
$$F_{A2} + C_1 W_{1,2} = F_{S2} + C_2 W_{1,2} \quad \text{(III)} \quad \text{(in ocean 2)}$$

$$0 = F_{A1} - F_{S1} + (C_2 - C_1) W_{1,2} \quad \text{(IV)}$$
$$0 = F_{A2} - F_{S2} + (C_1 - C_2) W_{1,2} \quad \text{(V)}$$

where C_1, C_2 are the concentrations of Nd in ocean 1 and 2, respectively.

putting ε Nd values in equation II yields:

$$\varepsilon_{A1} F_{A1} + \varepsilon_2 C_2 W_{1,2} = \varepsilon_1 F_{S1} + \varepsilon_1 C_1 W_{1,2}$$

where ε_{A1} : Nd isotopic ratio in continental source in A_1

$\varepsilon_1, \varepsilon_2$: Nd isotopic ratios in oceans 1 and 2 respectively.

These equations allow us to calculate the εNd value of ocean 1:

$$\left(\frac{\varepsilon_{A1} F_{A1} + \varepsilon_2 C_2 W_{1,2}}{F_{S1} + C_1 W_{1,2}}\right) = \varepsilon_1$$

ε values placed into equations IV and V:

$$0 = \varepsilon_{A1} F_{A1} - \varepsilon_1 F_{S1} + \varepsilon_2 C_2 W_{1,2} - \varepsilon_1 C_1 W_{1,2} \quad \text{(VI)}$$
$$0 = \varepsilon_{A2} F_{A2} - \varepsilon_2 F_{S2} - \varepsilon_2 C_2 W_{1,2} + \varepsilon_1 C_1 W_{1,2} \quad \text{(VII)}$$

Both ocean masses are completely decoupled from each other when W1, 2 (the rate of exchange of the water bodies) is equal to zero. In this case, FA1= FS1 and FA2 = FS2 are valid for the conditions of equilibrium in an ocean; the flux from continental source A1 corresponds to the sedimentation rate in ocean 1. Let us assume that the Nd concentrations in oceans 1 and 2 are the same (C1 = C2) and that the oceans are only very weakly linked to each other. For equations IV and V, this would yield:

$$(\varepsilon_1 - \varepsilon_{A_1})\, F_{S_1} \sim (\varepsilon_2 - \varepsilon_1)\, C_1\, W_{1,\,2} \qquad \text{(VIII)}$$

$$(\varepsilon_2 - \varepsilon_{A_2})\, F_{S_2} \sim (\varepsilon_1 - \varepsilon_2)\, C_1\, W_{1,\,2} \qquad \text{(IX)}$$

By placing the residence time τNd into equation VIII, the mixing time T between oceans 1 and 2 can be derived as follows:

For residence time τ:
$$\tau = \left(\frac{C_1 W_1}{F_{S_1}} \right) \qquad \text{(see equation I)}$$

where $W_1 = $ mass of water in ocean 1
and $F_{S_1} = F_{A_1} = $ flux of Nd into the ocean

(I) in (VIII):
$$\left(\frac{C_1 W_1}{\tau Nd} \right) \approx \left(\frac{\varepsilon_2 - \varepsilon_1}{\varepsilon_1 - \varepsilon_{A_1}} \right) C_1 W_{1,\,2} \qquad \text{(X)}$$

$$\left(\frac{W_1}{W_{1,\,2}} \right) \approx \left(\frac{\varepsilon_2 - \varepsilon_1}{\varepsilon_1 - \varepsilon_{A_1}} \right) \tau Nd \qquad \text{(XI)}$$

$W_{1,2} = W_1 / T$,   where W_1: mass of water in the ocean,
and T : rate of exchange

$T = W_1 / W_{1,2}$ (XII)

(XII) in (XI):
$$T = \left(\frac{W_1}{W_{1,\,2}} \right) \approx \left(\frac{\varepsilon_2 - \varepsilon_1}{\varepsilon_1 - \varepsilon_{A_1}} \right) \tau Nd \qquad \text{(XIII)}$$

ε_2 (Pacific) $\sim$ -3;
ε_1 (Atlantic) $\sim$ -12.0;
ε_{A_1} (Nd into the Atlantic) $\sim$ -14.4 (assuming this corresponds to North American shales "NASC")

$T = 4\,\tau Nd$; ($\tau Nd = 2000$ years)   (XIV)

$T = 8000$ years
============

Let us apply the lowest of the Nd residence times, 400 years, from Piepgras et al. (1979). Thus, 1600 years is shown to be the Atlantic-Pacific oceans' mixing time. Therefore, this lower residence time of 400 yr appears the most reasonable as 1600 years matches much better the calculated whole ocean mixing time (about 1500 years). Therefore, it is clear that the 'effectively dissolved' load of Nd and the REE in river water has been vastly underestimated. It is important to note the relationship between residence time and ocean mixing time in equation (XIV). From this relationship we can see that the mixing time must be about 4 times longer than the residence time of Nd.

These results explain the observation that there are different Nd isotopic ratios in the different ocean basins (Fig. 5.16). On the basis of the Nd residence time which must be shorter than the ocean mixing time, we can see that Nd can never be sufficiently circulated for isotopic homogenization similar to Sr. Piepgras and Wasserburg (1980) demonstrated that inhomogeneities in the Nd isotopic ratio can exist even within one ocean basin. Deep water samples from the Atlantic (>1000 m depth) show relatively homogeneous εNd values of -13.5 ± 0.4. Samples from shallower depths, however, are far more radiogenic (less continental influence) with εNd values of between -9.6 and -10.9. Similar variations in both the Atlantic and the Pacific could be observed by Piepgras and Wasserburg (1983) (Fig. 5.17).

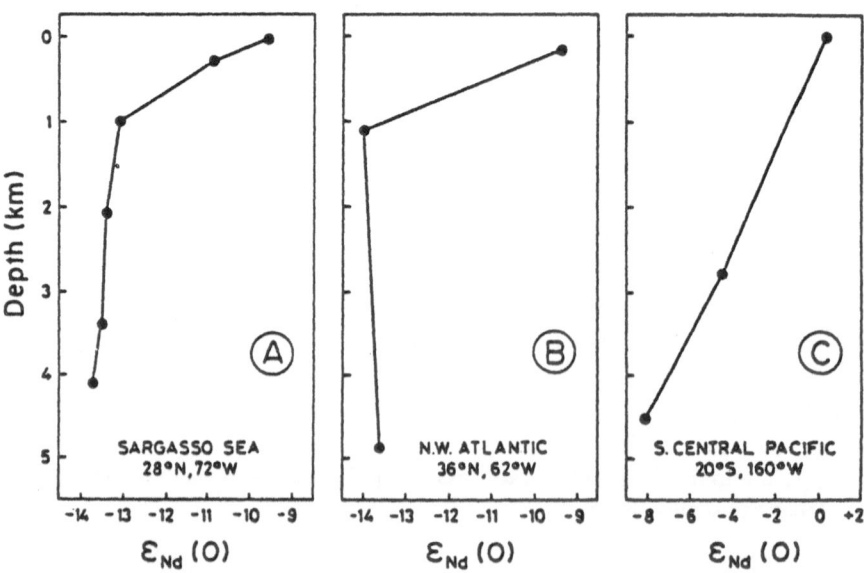

Fig. 5.17. Variation of the Nd isotopic composition within three marine basins as a function of water depth. (Piepgras and Wasserburg 1983)

As a consequence of such variation, Nd isotopes have great potential for oceanographic investigations as they allow us to differentiate several water bodies, to recognize ocean currents, and to reconstruct exchange processes within a single ocean basin (Sect. 5.3).

5.2.2 Possibilities for the Derivation of the Nd Isotopic Composition of Paleo-Oceans

The first studies which looked at the Nd isotopic composition of paleo-oceans were carried out on metal-rich sediments, Mn-crusts and nodules. As already mentioned, these preferentially incorporate Nd, which retains the isotopic composition of seawater. Worth mentioning in this connection are the studies of Chyi et al. (1984) and Hooker et al. (1981). Analogous to studies on recent metal-rich sediments, Mn crusts and ores, they investigated ancient marine ore bodies with known ages and so, for the first time, were able to derive the Nd isotopic compositions of Jurassic and Cambrian seawater. Their investigations also involved marine carbonates, phosphates and cherts.

Shaw and Wasserburg (1985) carried out Nd isotopic analyses on recent carbonates and phosphates from the Atlantic and the Pacific oceans and showed that these phases also incorporate Nd with a seawater signature. The Sr and Nd isotope ratios of modern day carbonates and phosphates are consistent with those of today's oceans. The isotopic analyses of fossil carbonates and phosphates of Shaw and Wasserburg (1985) provided information about the isotopic evolution of paleo-oceans for the first time. Their εNd values are shown in Fig. 5.18, where

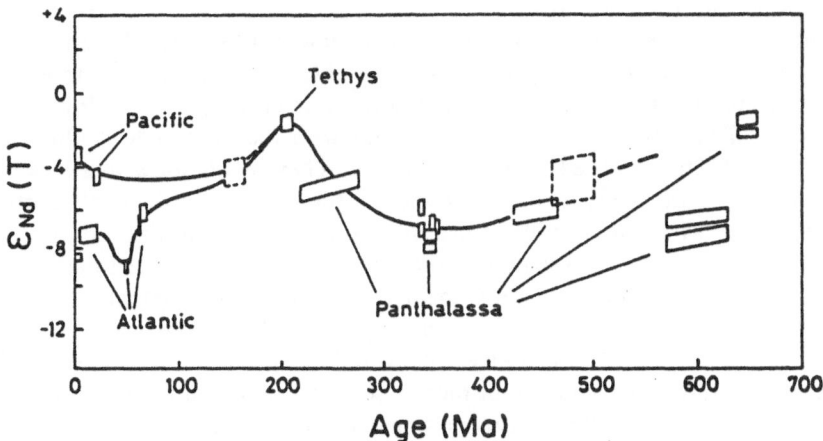

Fig. 5.18. Nd isotopic variations in seawater over the last 600 Ma. (Shaw and Wasserburg 1985)

these data and compared with those from Fe-Mn-rich sediments investigated by Hooker et al. (1981) and Chyi et al.(1982). Independently varying Nd isotope evolution curves can be observed for the last 200 million years. 200 Ma ago, at the opening of the central Atlantic, both oceans, the Pacific (Panthalassa and the Paleotethys) and the Atlantic had the same isotopic characteristics. At that time, the 'super ocean' Panthalassa, which surrounded the supercontinent Pangaea, may still have existed. The further opening of the Atlantic Ocean went hand in hand with the development of its unique isotopic signature. This means that the Pacific and the Atlantic received their rare earth element intake from isotopically different sources. These observations made it clear for the first time that, unlike Sr, the Nd isotope system is particularly well suited to support paleo-oceanographic as well as paleo-tectonic models. We will come back to this later on. In the next part, we discuss the suitability of carbonate and phosphate material for the derivation of Nd isotopic composition in the paleo-oceans.

Carbonates

The Nd concentrations in modern biogenic calcites and aragonites are low and vary between 0.2 and 9.6 ppb. Inorganic aragonite precipitate can have concentrations of up to 65 ppb. Limestones and dolomites likewise show low Nd concentrations of a few hundred ppb. The Sm/Nd ratios of carbonates vary stronger than those in seawater and allow us to surmise that there must be some fractionation during the incorporation of the REE. Banner et al. (1988) point out that the REE can be mobilized and fractionated during carbonate diagenesis. In contrast to recent carbonates, carbonate fossils show far higher Nd concentrations (1-30 ppm). They are therefore more greatly enriched in Nd by a factor of 10^3-10^4 over recent carbonate bivalves.

The observations of Turekian et al. (1973) allow us to explain this enrichment process. These authors were the first to be able to show that the rare earth elements are preferentially incorporated into sub-microscopic Fe-hydroxides on Fe-flakes. These flakes are deposited along with the carbonate. Palmer and Elderfield (1986) observed something similar. They could show that less than 10% of the REE were actually found in the calcite lattices of Tertiary foraminifera. More than 90% were found in Fe-Mn oxides coating the foraminifera skeletons. These crusts build up after the death of the foram during the sedimentation process. This means that the REE contents of recent foraminifera are determined by the compositions of these crusts and not by the calcite of the foraminifera themselves. Yet still these recent foraminifera show Nd isotope signatures characteristic of today's seawater. How do fossil foraminifera behave? Can they retain the Nd isotope signatures of ancient oceans?

Palmer and Elderfield (1986) showed that Mn/Fe ratios in Tertiary foraminifera are far higher than in foraminifera today. This implies either that the accumulation rates of Mn and Fe have changed over time or that the Mn and Fe crusts have been altered during diagenesis. Although changes in accumulation rates are possible, it appears more likely that the Mn/Fe ratios have been altered during diagenesis. Froehlich et al. (1979) showed that Mn and Fe are very mobile during anoxic diagenesis. As the concentrations of the REE in recent samples can be correlated with Fe content, it can be assumed that the transformation of an Fe-rich phase with low Mn/Fe ratios to a Mn-rich phase with high Mn/Fe ratios must lead to remobilization of the REE. It is possible that the Fe-rich crusts of recent foraminifera are dissolved during this kind of diagenetic process leading to a redistribution of the REE. High Mn/Fe ratios can therefore be used as an indicator of diagenetic change. In Fig. 5.19, the REE distribution in recent foraminifera is compared with that for the Tertiary. Clear differences can be seen in these patterns. Particularly striking is the characteristic negative cerium (Ce) anomaly, which is found in the recent samples but is seen much less clearly in those from the Tertiary. On the basis of these observations, it can be assumed that REE distributions can be strongly modified by diagenetic processes, not only in fossil foramifera but also in other carbonate-rich sediments.

These investigations show that Nd isotopic data from carbonates may only be used with care for the derivation of primary seawater Nd isotopic composition. It is only to be expected that some diagenetic fluid phases which are not in equilibrium with seawater may overprint the isotopic composition of lithified carbonates. In the case of these foraminifera, Palmer and Elderfield assume that the Fe-Mn crust formation took place during an early sub-oxic phase of diagenesis, which was still strongly influenced by seawater. In this case, isotopic analyses are still likely to yield information about the paleo-ocean.

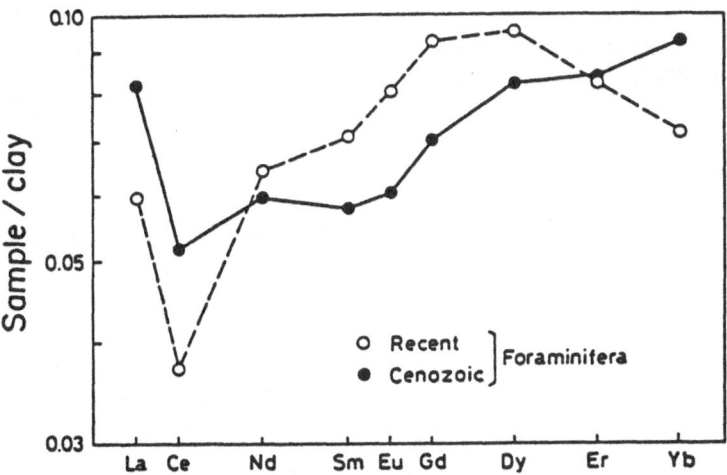

Fig. 5.19. Shale normalized REE patterns in foraminifera. Differences between recent and fossil foraminifera from the Cenozoic manifest themselves in a weakening of the Ce anomaly. (Palmer and Elderfield 1986)

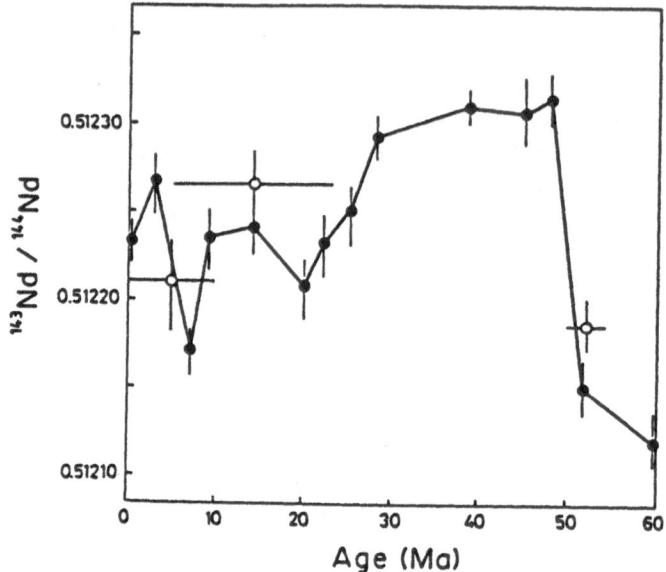

Fig. 5.20. Nd isotopic variations of Atlantic seawater over the last 60 Ma reconstructed with the help of isotopic analyses of foraminifera (Palmer and Elderfield 1986). Open circles are data from Shaw and Wasserburg (1985).

The isotopic analyses carried out on these foraminifera allow the reconstruction of the evolution of the Nd isotopic composition of Atlantic seawater over the last 60 million years (Fig. 5.20). Three data points from the study of Shaw and Wasserburg (1985) are also shown. They fit in well with the data of Palmer and Elderfield (1986).

Phosphates

Modern, biogenic phosphate is also characterized by very low concentrations of the rare earth elements (Shaw und Wasserburg, 1985). Thus, shark teeth, phosphatic brachiopods, and fish bones all contain between 5 and 150 ppb. Fossil biogenic apatite is likely to contain up to 10^4 times higher Nd concentrations. Ancient biogenic and inorganic apatite in phosphorite deposits is also strongly enriched in the REE. Investigations show that biogenic phosphate incorporates REE directly after the death of the organisms. In contrast to the carbonates, phosphates actually incorporate the REE directly into the crystal lattice as a result of Ca substitution ($REE^{3+} \longleftrightarrow Ca^{2+}$). Therefore, although recent phosphate contains Sr and Nd whose isotopic compositions reflect those in seawater it can not be immediately assumed that fossil phosphates will yield a seawater isotopic signature.

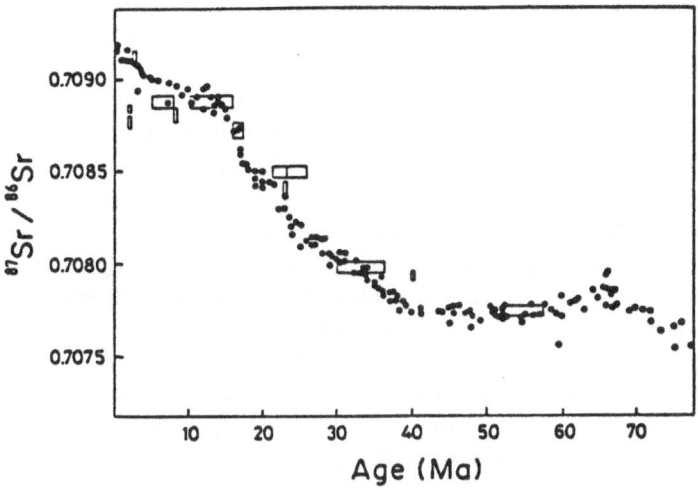

Fig. 5.21. Sr isotopic ratios determined from fish teeth phosphate. These data fit on the trend of seawater Sr isotope evolution for the last 70 Ma. (Staudigel et al. 1985)

Staudigel et al. (1985) investigated phosphate from fossil fish teeth. The initial Sr isotopic compositions of the phosphate fit in well with what we know about the evolution of seawater Sr isotope ratio over the last 70 million years (Fig. 5.21). This correspondence implies that the isotopic ratios of this Sr and possibly Nd have undergone negligible diagenetic alteration.

In Fig. 5.22 initial Nd isotopic ratios are plotted against stratigraphic age. The samples show a region of variation in isotopic composition which could already be made out in the study of Shaw and Wasserburg (1985). The Nd isotopic compositions of the Pacific and the Atlantic can easily be differentiated from each other for the last 80 Ma. As recent fish phosphate samples show the same Nd isotopic compositions as in seawater and in manganese nodules, the authors assume that the REE in fossil fish teeth likewise derive from seawater. However, using Nd isotopic compositions of organic phosphates for the reconstruction of trends in seawater Nd isotope ratio must be carried out with caution. As we have already observed, recent fish bones are characterized by relatively low Nd concentrations. After the death of a fish, this organic phosphate becomes enriched in REE. We can scarcely expect that these REE will always derive from seawater without exception.

Grandjean et al. (1987) also considered that Fe-Mn oxihydroxide flakes in seawater are responsible for the enrichment of REE after deposition. These flakes exist in large quantities in the ocean and scavenge REE directly from seawater. The flakes end up eventually in the sediment. The reducing conditions in the sediment result in the dissolution of these Fe-Mn hydroxides and make it possible for the transfer of REE into the phosphate during the transformation of organic phosphate to carbonate fluoro-apatite, the most common form of rock phosphate.

In this case, the isotope ratios and the REE patterns should reflect those of seawater. The high REE concentrations that appear even in early phases of diagenesis have the consequence that the phosphate may remain relatively insensitive even to later diagenetic exchange processes and are likely to retain their original isotopic composition. Such considerations argue very much in favour of using phosphatic rocks or fossils for Nd isotopic work.

Grandjean et al.'s arguments concerning REE enrichment close to the sediment/seawater interface are illuminating, although the isotopic data allow us to establish that not only marine Sr and Nd were incorporated into the phosphate lattice during diagenetic alteration. Although Sr isotopic ratios for some samples do lie on or near the Sr evolution curve, some quite clearly do not.

The recent studies of Stille et al. (1994; 1997) und Riggs et al. (1997) on various types of marine phosphate material (brachiopods, vertebrae, teeth, bone fragments and peloidal grains) showed that phosphatic peloids are by far the most suitable phosphatic material for the determination of seawater Sr and Nd isotopic composition.

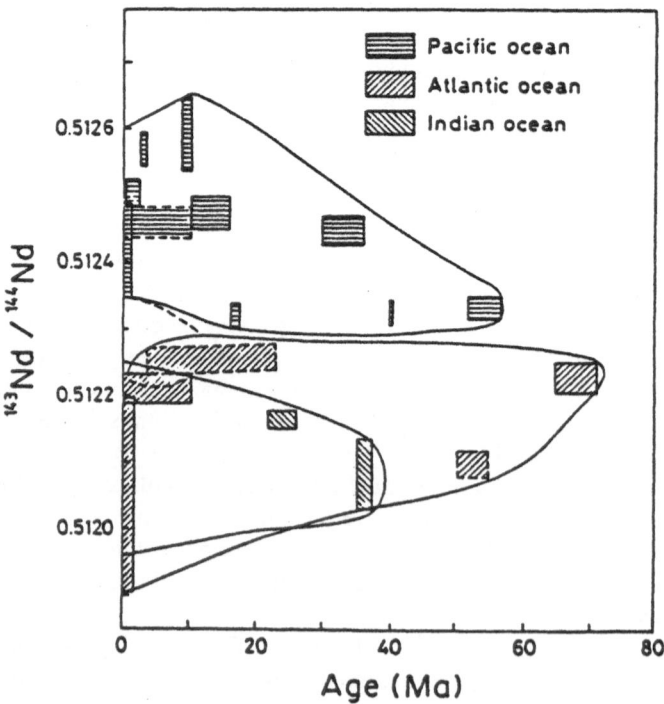

Fig. 5.22. Nd isotopic ratios determined on fish teeth phosphate. The Nd isotopic ratios of the Atlantic and Pacific oceans must have been quite different over the last 70 Ma. (Staudigel et al. 1985)

Fish teeth and bones can vary strongly showing isotopic ratios which can be far removed from that of coeval seawater. This would imply that their diagenetic systems are more open to exchange rendering them more susceptible to contamination from non-seawater sources.

In order to derive seawater Nd isotopic composition using phosphates, it is important that a control be first carried out by measuring the Sr isotopic composition of the material and comparing this with the established curve for seawater Sr isotope evolution. Only samples yielding primary Sr isotopic compositions and seawater REE patterns (HREE enrichment and Ce anomaly) should then be used to derive the seawater Nd isotopic composition.

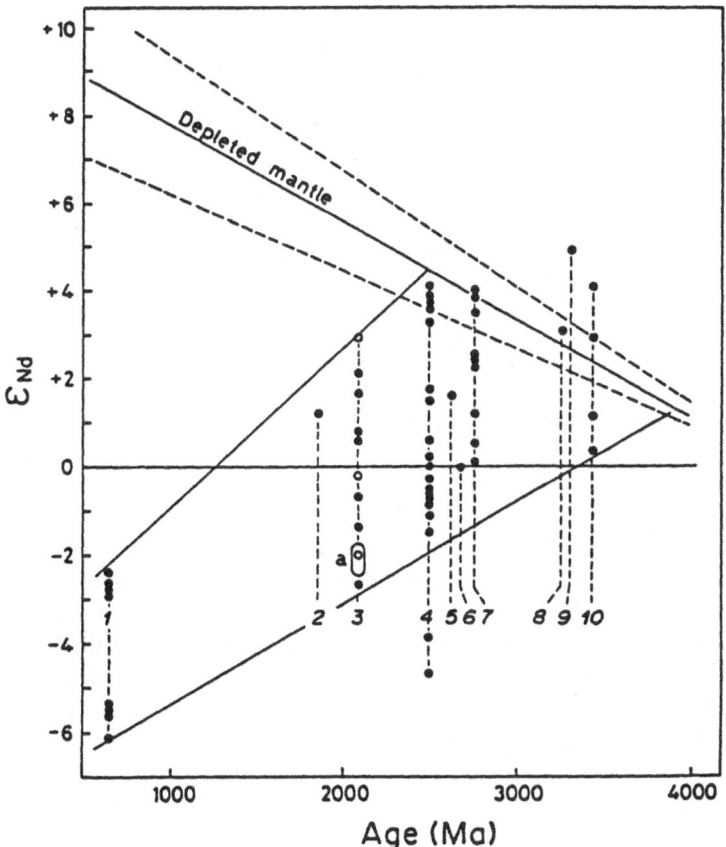

Fig. 5.23. Possible Nd isotopic composition of Precambrian seawater based on measurements of 'Banded Iron Formations' (full circles) of 1) Urucum, 2) Sokoman, 3) Gunflint, 4) Hammersley, 5) Vermillion, 6) Bjornevann, 7) Michipicoten, 8) Beartooth, 9) Cleaverville and, 10) Fig Tree. Open circles represent phosphates from Gabon. Authigenic clays from the Gunflint formation are found in the region 3a. Data from Miller and O'Nions (1985), Stille and Clauer (1986), Jacobsen and Pimentel-Klose (1988), Bros (1993).

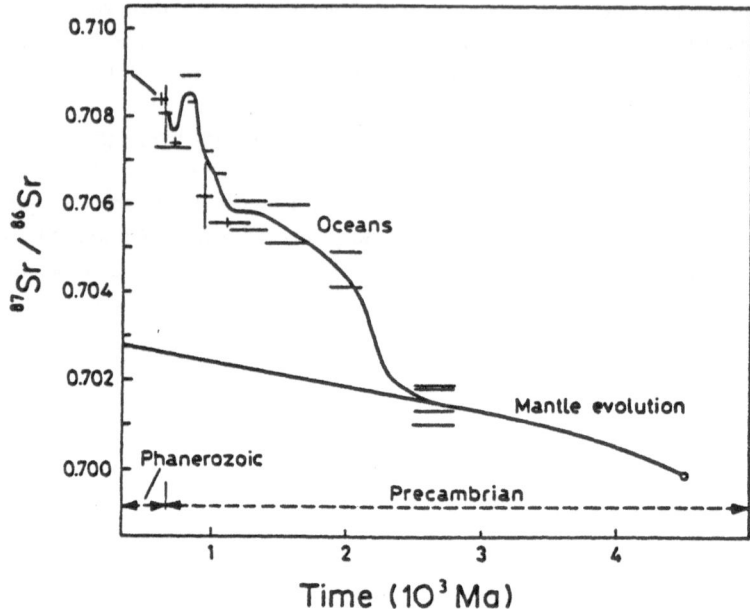

Fig. 5.24. The Sr isotopic composition of Archean-Proterozoic seawater. (after Veizer and Compston 1974; 1976; Burke et al. 1982)

5.2.3 Nd Isotopic Evolution in Seawater

Precambrian and the Paleozoic

There are few available data for the reconstruction of isotopic evolution of Nd in the Precambrian. There are some data from iron-rich, Precambrian sediments which have been placed under the general title of "Banded Iron Formations" or "BIFs". These sediments were deposited widely during most of the Precambrian and can be found today in rocks from the Americas, Australia, Africa and northern Europe. It has generally been assumed that these sediments are of marine origin and it has been maintained that these sediments have incorporated the Nd isotopic composition of coeval seawater. Miller and O'Nions (1985), Stille and Clauer (1986), and Jacobsen and Pimentel-Klose (1988a,1988b) were the first to investigate these sediments isotopically.

They report strong variations in initial εNd values of more than 6ε units. Such strong variations in initial Nd isotopic composition, found in samples of about the same age could be the result of the presence of detritus. This seems all the more likely if we consider that there are no mineralogic or X-ray diffraction data for many of these samples. The initial εNd rise begins in the region of mantle values at about 2500 Ma. This trend implies that early Precambrian seawater, in particular that of the Archean, was more strongly influenced by the mantle than seawater of the later Precambrian or the Phanerozoic. It remains, however, unclear to what extent erosion of isotopically variable continental crust (with mantle isotopic signatures) or enhanced hydothermal activity may have influenced the isotopic signature of Archean seawater. The Sr isotopic system was not yet decoupled from that of the Nd isotopic system and yields isotopically identical information (Fig. 5.24)

Palaeozoic

Hooker et al. (1981) were the first to investigate the Nd isotopic composition of the Paleozoic Iapetus ocean basin and attempted this using metal-rich sediments. Investigations of Keto und Jacobsen (1987, 1988) on phosphatic brachiopods and conodonts from Europe und North America followed. They were able to show that Paleozoic seawater displayed similarly strongly variable Nd isotopic ratios (Fig. 5.25) and discovered that for the time period between 600 and 400 Ma there were two paleogeographically independent marine isotope trends. The authors were able to show for the first time that the Iapetus Ocean, which was just starting to develop in the latest Precambrian, possessed a different and more radiogenic isotopic composition. They surmised that the Iapetus Ocean was separated from the Panthalassa Ocean by a physical barrier of some kind.

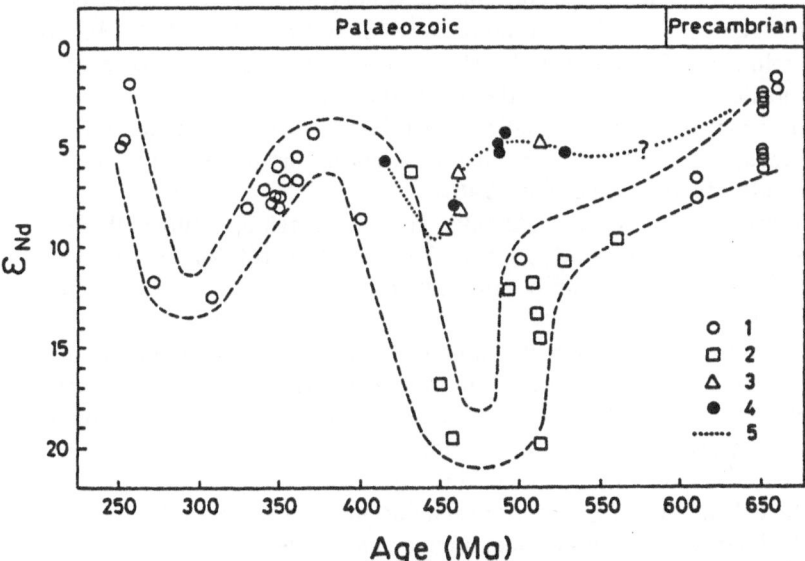

Fig. 5.25. Nd isotopic composition of seawater during the Paleozoic (Keto and Jacobsen 1987, 1988). The Nd isotopes tell us that the Iapetus Ocean with its more radiogenic isotopic signature must have been decoupled from the Panthalassa ocean. Samples 1 and 2: North America. 3: SE USA. 4 Samples from Europe. 5: Iapetus ocean.

5.2.4 The Closing of the Tethys and the Opening of the Atlantic

Mesozoic and Cenozoic

The present day oceans formed during this period. It is thus of great paleogeographic interest to find out how the isotopic compositions of the Pacific, North and South Atlantic, Indian and Tethys Oceans and also the Mediterranean developed through the Mesozoic and Cenozoic. Variations in the Nd isotopic ratio should reflect the changing interlinked relationships between these oceans (Fig. 5.26). In order to derive this Nd isotope evolution, not only phosphate but also carbonate was used whose stratigraphic age is known and whose initial Sr isotope ratios are identical to seawater at the time of formation (see data summary in Stille 1992; Stille et al. 1996).

The Nd isotopic composition in the Pacific has remained almost constant over the last 200 million years and shows εNd values that vary between -5 und -1. The early Tethys seawater shows identical isotopic ratios to those of the Pacific which implies that Pacific seawater found its way into the newly forming Tethys basin between 200 and 180 Ma. The influence of continental Nd on the Nd isotopic composition of the Tethys, by way of rivers and resedimentation along the rift zones became important very soon after the splitting up of the Tethys basin and caused a lowering of the Nd isotope ratio. To what extent this was brought about by the development of a physical barrier is not clear. The Nd isotopic composition of the Tethys became ever more decoupled from that of the Pacific. This decoupling was first observed by Shaw and Wasserburg (1985). Measurements on phosphatic concretions from the Vocontian Trough, which was located in the northern part of the Tethys indicate that Tethys seawater reached lowest, almost continental crust-like εNd isotope values during the late Early Cretaceous some 100 to 120 Ma ago (Stille et al. 1996).

No data exist for the early Atlantic. However it can be assumed that its isotopic composition was the same as for the Tethys before 100 Ma because there was a wide seaway between both oceans at that time and ocean currents were directed East-West.

Let us take a look at the evolution of Nd isotopic composition in the Atlantic and the Tethys over the last 80 million years. This was a time interval during which significant tectonic events took place (e.g. the rifting open of the South Atlantic, Himalayan mountain building). Fig. 5.27 illustrates their two very different behaviors of the Sr and Nd isotopic systems in seawater. Between 60 and 25 Ma, Nd and Sr isotopic ratios rose in the Atlantic and in the Tethys. From 25 to 20 Ma, both isotopic systems measured on carbonates and phosphates from North and South Atlantic, and Tethys oceans provide the same information with Sr isotope ratios rising and Nd isotope ratios falling, impling an increase in crustal influence. The most striking aspect of this diagram is the sudden and rapid rise in Nd isotopic composition around 20 Ma in the Atlantic ocean. This rise does not manifest itself in the Tethys where the Nd system remained strongly influenced by continental input. This jump takes place without any reaction from the Sr system and thus cannot be related to a sudden switch in input from continental to basaltic, oceanic sources. Such a switch would not have led to a decoupling of the Sr and the Nd systems. That is to say, an increase in the amount of mantle influenced Sr and Nd would have led to a rise in the Nd isotopic compositions coupled with a fall in Sr isotope composition. This is not the case here.

Stille (1992) suggested that these Nd anomalies (decouplings) at 20 Ma and between 60-25 Ma are the consequence of large-scale tectonic events which altered the paleo-currents in the oceans. In this respect, three aspects are of importance:

1) The Sr isotopic composition is, and was, homogeneous in the oceans.

2) The various oceans show different Nd isotopic compositions.

3) The Pacific shows a consistently more radiogenic Nd isotopic signature than the Tethys and the Atlantic over the last 100 Ma.

A change in circulation, i.e. communication between the oceans, can be recognized in the Nd isotope composition of oceans without any subsequent change in the Sr isotope composition of that seawater. The changes in the isotopic evolution of the Tethys and the Atlantic over the last 100 million years can be explained as follows.

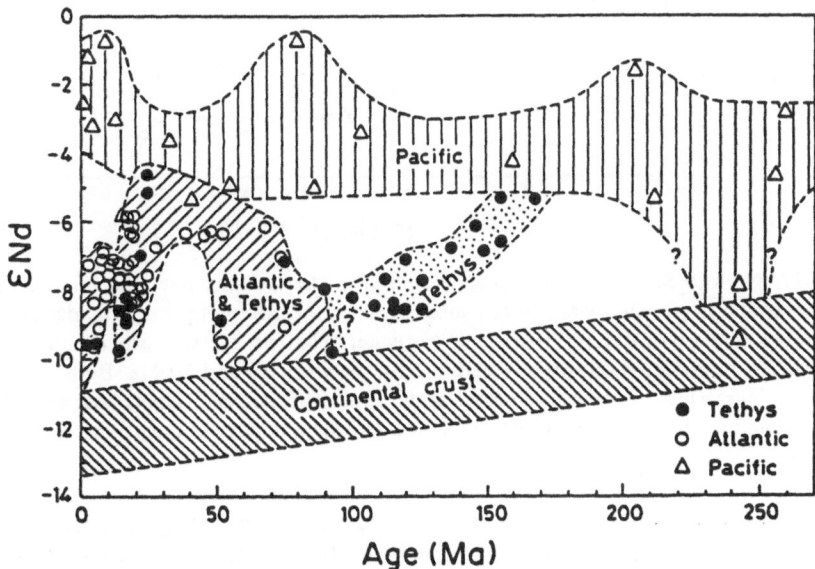

Fig. 5.26. The Nd isotopic evolution in the Pacific, Atlantic and Tethys over the last 240 Ma. Variations in the Nd isotopic ratios reflect the changing relationships between these oceans. (Stille et al. 1996)

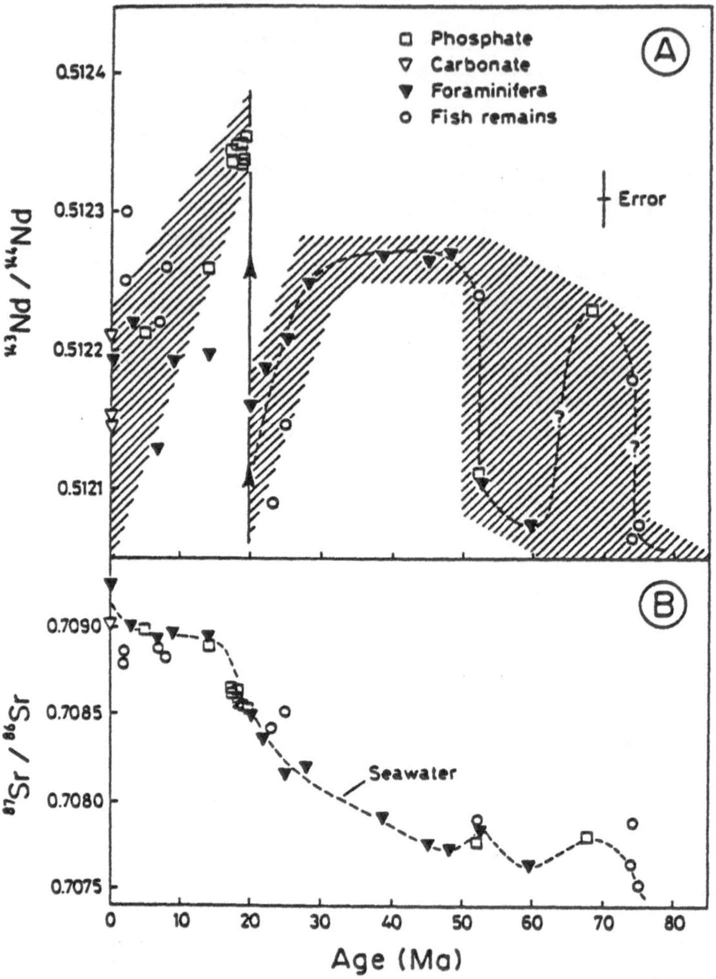

Fig. 5.27. Nd and Sr seawater isotopic evolution in the Atlantic over the last 80 Ma. Comparing the two evolutionary trends sheds light on the differently reacting, decoupled behaviours of the two isotopic systems in seawater. The two systems provide the same geochemical information during one period only: between 30 and 20 Ma. Both systems respond to increasing continental influence at this time. The anomalous behaviour at 20 Ma can be related to plate tectonic events and their subsequent impact on the paleo-currents in the oceans. (Stille 1992)

With the opening of the South Atlantic and the northward drifing of India around 100-80 Ma, a new seaway developed, which allowed the westwardly directed Pacific currents to cross the Indian ocean and end up in the basins of the S. and, eventually, the N. Atlantic (Fig. 5.28). Pacific seawater with a significantly higher Nd isotopic composition began to mix progressively with the waters of the Atlantic, which led to the rise in Nd isotope ratios between 60 and 50 Ma.

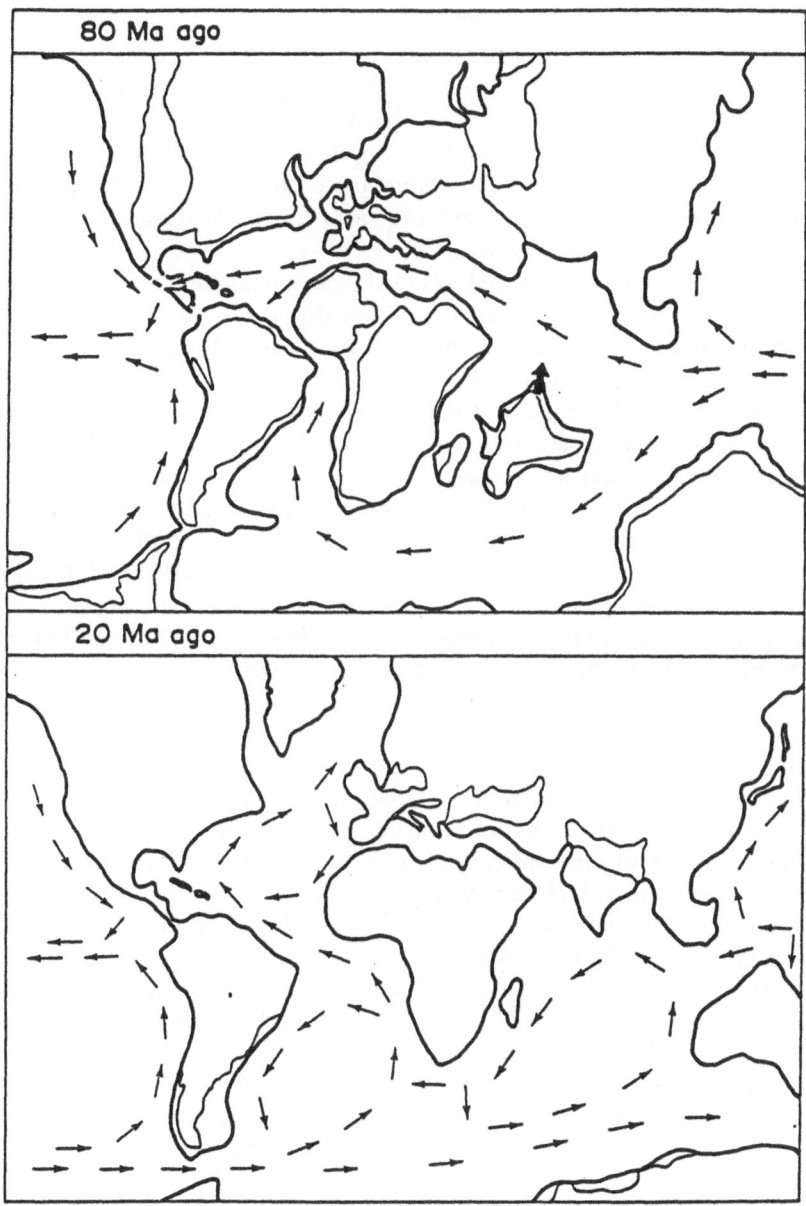

Fig. 5.28. The connecting routes and currents of communicating ocean water bodies 80 and 20 Ma respectively.

In the case of Tethys seawater we observe an even stronger increase in the Nd isotopic ratio reaching values at 25 Ma (Late Oligocene/Early Miocene) even higher than those observed for contemporaneous Atlantic seawater. This indicates that at this period larger masses of Pacific surface waters entered again directly through the Indian-Tethys seaway into the Tethys basin. Stille et al. (1996) suggest that this increased inflow of Pacific seawater into the Tethys was probably directly related to a first order sea level highstand some 80 to 50 Ma ago (Late Cretaceous/Early Tertiary). This highstand could apparently completely compensate for the plate tectonic induced narrowing of the Tethys seaway which also started in the Late Cretaceous.

In the early Miocene (25 Ma) the Nd isotopic composition of Tethys seawater was intermediate between Atlantic and Pacific seawater. This supports the hypothesis, that in spite of the narrow seaway it was still essentially Pacific and not Atlantic seawater which fed the Tethys basin.

The significant decrease of Nd isotopic compositions observed in South Atlantic, North Atlantic and Tethys seawater indicates that around 25 to 20 Ma much less Pacific seawater entered these oceans. The strongest decrease is observable for Tethys seawater. This change of Nd isotopic composition is a hint that with the collision of the Arabian and Eurasian plates the Indian-Tethys seaway started to close some 25 Ma ago leading to the breakdown of the circum-equatorial circulation patterns of the world's oceans some 20 Ma ago.

A strong decrease of the Nd isotopic composition around 25 to 20 Ma is also observable for the North and South Atlantic. It is, however, not probable that it was the closing of the small Indian Tethys seaway alone which caused the strong shift. Stille et al. (1996) suggest that it was also due to less radiogenic seawater which entered the Atlantic from the South. A possible explanation for this could be the separation of Australia from Antarctica about 45 Ma ago and its northward drift and collision with Indonesia during Middle Miocene times. This plate tectonic event led to the evolution of today's circulation pattern in the Pacific realm with two individual current systems for the Indian and Pacific oceans. Consequently, it was no longer simply Pacific, but Indian ocean water that entered the Atlantic from the South around Africa 23 to 20 Ma.

At about 20 to 18 Ma the Nd isotopic compositions of the Atlantic and Tethys oceans evolved differently. Whereas the Nd isotopic composition of Tethys seawater continued to drop, the isotopic composition of North Atlantic seawater started to increase again. This decoupling of the Nd isotopic systems of the Tethys and the Atlantic oceans indicates a narrowing of the seaway between both oceans at Gibraltar. This led to an almost complete isolation of the Tethys which was now becoming the present-day Mediterranean sea.

The strong increase of the Nd isotopic composition in the Atlantic might have been caused by major changes in circulation patterns within the Atlantic ocean itself some 25 to 20 Ma. They might have been mainly triggered by the opening of the Drake passage to the South of South America which led to the establishment of the circum-Antarctic current system. These currents enabled Pacific seawater to flow around Antarctica and enter the South Atlantic through the Drake passage and to flow northward to join the Gulf Stream in the Gulf of Mexico. This new admixture of more radiogenic Pacific seawater could explain the increase of the Nd isotopic composition in the Atlantic some 20 to 18 Ma ago (Stille 1992; Stille et al. 1996).

18 to 15 Ma the Nd isotopic composition of the Atlantic ocean dropped again and evolved to present-day seawater values (Fig. 5.26). This might indicate that the influx of Pacific seawater into the Atlantic diminished very soon after the opening of the Drake Passage. However, Stille (1992) and Stille et al. (1996) suggest that this change in Nd isotopic composition is directly related to the development of North Atlantic Bottom Water some 15 Ma ago. No data exist for the isotopic composition of the paleo-North Atlantic bottom water. However, investigations of present-day North Atlantic Bottom Water demonstrate that they are less radiogenic in Nd, and have the lowest Nd isotopic compositions ever measured in the open oceans (Piepgras and Wasserburg 1983). Because upwelling of cold, nutrient-enriched bottom waters became a widespread phenomenon during the Miocene leading to the deposition of primary phosphate sediments (Riggs, 1984), it has been suggested that exchange of these non-radiogenic bottom waters with surface waters caused the decreasing isotopic composition values in the Atlantic ocean (Stille 1992; Stille et al. 1996).

The short residence time of Nd clearly provides great potential for identifying changes in ocean circulation and can be of help in reconstructing paleogeography. However, equally rewarding aspects of this isotopic system may rest in its coupling with other geochemical parameters.

5.3 Tracing Ocean Currents Using Nd and Pb Isotopes

5.3.1 Ocean Currents and Nd Isotopes

In the study of Piepgras and Wasserburg (1983) seawater exchange was investigated between the Mediterranean and the Atlantic near the Strait of Gibraltar with the help of Nd isotopes. The outflow from the Mediterranean into the Atlantic has a significant influence on the temperature and the salinity of the water in the Atlantic. The high rate of evaporation in the Mediterranean Sea produces water with salinities greater than 38 per mil. This dense, salty water flows past Gibraltar at 10^6 m^3/s into the Atlantic. At the same time, less saline water flows from the surface waters of the Atlantic at the nearly equivalent rate of 1.11×10^6 m^3/s into the Mediterranean. The higher density of Mediterranean water sinks and flows at depths of more than 1000 meters along the continental shelf of the Atlantic.

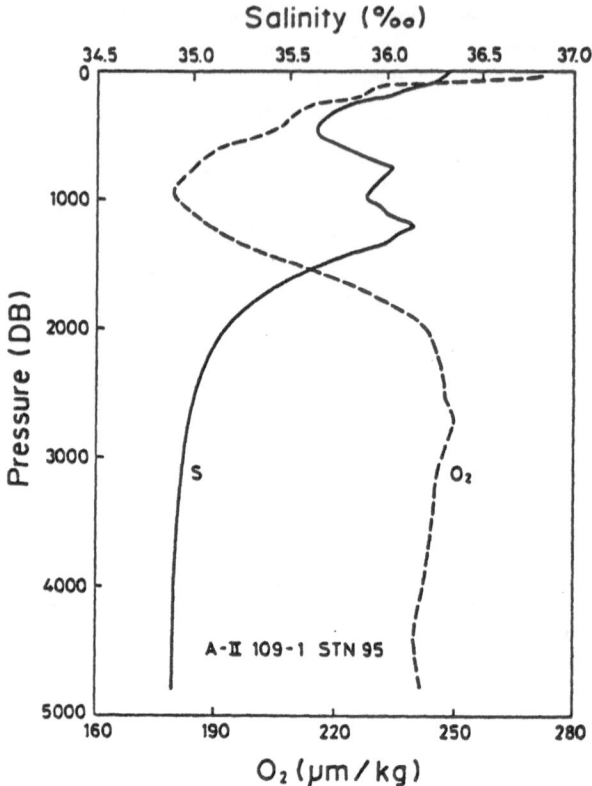

Fig. 5.29. Relationship between salinity, oxygen content and water depth at Gibraltar. (Piepgras and Wasserburg 1983)

It can be assumed that the flow of water from the Atlantic into the Mediterranean greatly influences the isotopic composition of Mediterranean seawater. The changes in salinity and oxygen content with water depth can be seen in Fig. 5.29. The salinity drops from 36.5 to 35.6 with increasing depth from surface waters down to depths of 500 m. At 500 m, salinity rises again to values around 36.1 by 1200 m depth and then decreases again to around 25 per mil. Oxygen content behaves in exactly the opposite fashion. The lowest observed oxygen contents and highest salinities can be correlated with the horizon of flow of Mediterranean waters into the Atlantic. The εNd values vary over the whole water profile by about 2.7 ε units (Fig. 5.30). The uppermost 500 m of the water column show εNd values between -11.5 und -12.5. At depths greater than 500 m, these values rise reaching maximum values of around -9.8 +/- 0.6 at about 1000 m depth. At this depth, the εNd values fall again. The Sm und Nd concentrations are likewise dependent on water depth and rise with increasing depth (Fig. 5.31). The maximum εNd values between depths of 800 and 1000 m can easily be differentiated from the other Nd isotopic ratios of the water profile.

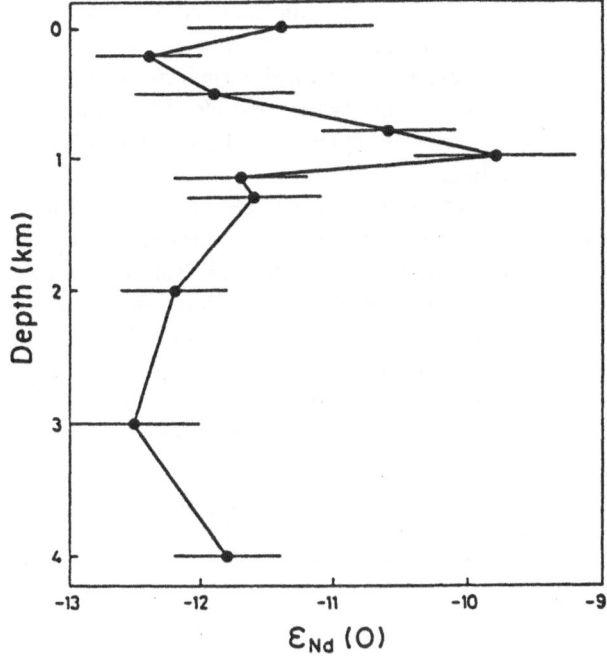

Fig. 5.30. Nd isotopic variation as a function of depth at Gibraltar. The exchange between the water bodies of the Mediterranean and the Atlantic can be shown with the help of Nd isotopes. (Piepgras and Wasserburg 1983)

The difference allows us to surmise that this Nd must be derived from a source other than that for the waters above and below this layer. The maximum value lies 2ε units from that of the other water samples and implies that Mediterranean water with a εNd value of about -9.8 has flowed into the Atlantic, which has εNd value of about -12. A εNd value of -9.8 makes it clear that some other source with much higher (more mantle influenced) εNd than the Atlantic must supply the Mediterranean with REE (Frost et al. 1986).

5.3.2 Ocean Currents and Pb Isotopes

Pb concentrations in marine bottom waters are also vanishingly small: around 1 ng/L (Flegal et al. 1986, 1989; Flegal and Patterson 1983). This is what makes Pb such a sensitive tracer of anthropogenic contamination in the environment (see Sect. 4.1; such widespread contamination is a cause for concern in every radiogenic isotope laboratory!). The residence time for Pb in seawater is on the order of 80 to 100 years and so is even shorter than that of Nd (Craig et al. 1973). We would therefore expect that the Pb isotopic ratios vary from ocean to ocean as in the case of Nd. Chow and Patterson were already able to show in 1962 that such differences in Pb isotopic compositions do exist between ocean basins.

By virtue of the short residence time of Pb in seawater, the Pb isotope system ought to react even more sensitively to local variations in Pb input through space and time than does the Nd isotope system for Nd. Pb isotopes also represent a further opportunity to trace ocean currents and the extent of exchange between various oceans through time.

The direct measurement of Pb isotope ratio in natural fluids is extremely difficult to perform as Pb concentrations are generally very low indeed. Until today, scarcely one direct analysis of seawater Pb isotope ratios has been carried out successfully and published. Nevertheless, Chow and Patterson (1962) were able to show that marine authigenic Mn ores incorporate seawater Pb isotopic ratios. Although this pioneering study showed the way, it was not until the recent studies of Abouchami and Goldstein (1995) that the behaviour of Pb isotopes in the marine environment began to be investigated using manganese nodules.

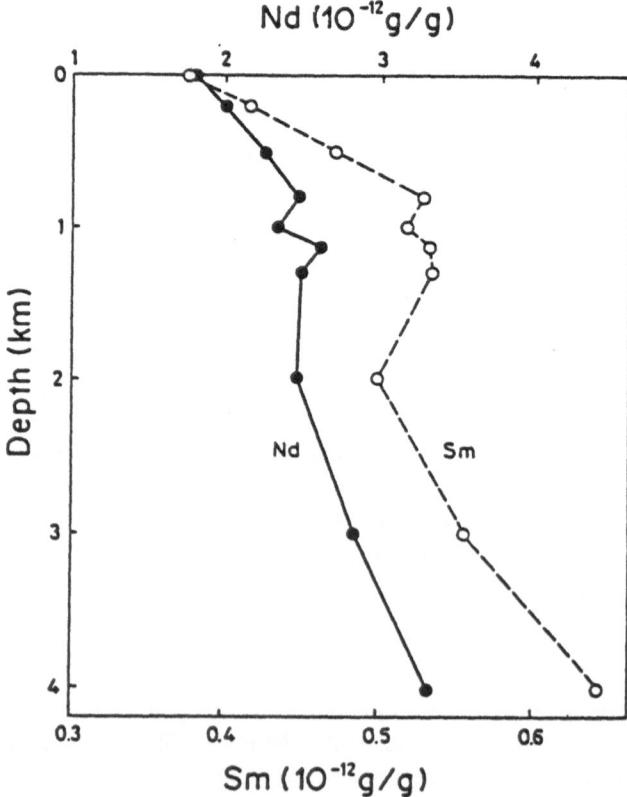

Fig. 5.31. The Sm-Nd concentrations change with water depth at Gibraltar. (Piepgras and Wasserburg 1983)

Nd and Pb isotope determinations of more or less recent, hydrogenous iron-manganese crusts have shown that their isotopic ratios have varied as a consequence of today's deep ocean currents (Albarède and Goldstein 1992; Abouchami and Goldstein 1995). The Nd isotopic ratios reflect not only these deep sea currents but also the geology of the sea floor itself. In the Atlantic, maximum εNd values are characteristic of the region of the mid-oceanic ridge, which differentiates the western and eastern deep sea basins from each other.

The work of Abouchami und Goldstein (1995) showed clearly that Pb isotopes reflects the changing behaviour of ocean currents far more sensitively than Nd isotopes by virtue of the short residence time of Pb in seawater. The authors investigated manganese nodules from the circum-Antarctic Ocean (Fig. 5.32).

The Pb isotope ratios of these nodules varied systematically with the geographic location of the samples (Fig. 5.33). The $^{206}Pb/^{204}Pb$ ratio drops gradually as one moves eastward in the A-Pacific (A=Antarctic). This isotopic evolution permits the assumption that greater influence from the Pacific can be related to lower $^{206}Pb/^{204}Pb$ ratios in the eastward flowing circumpolar water.

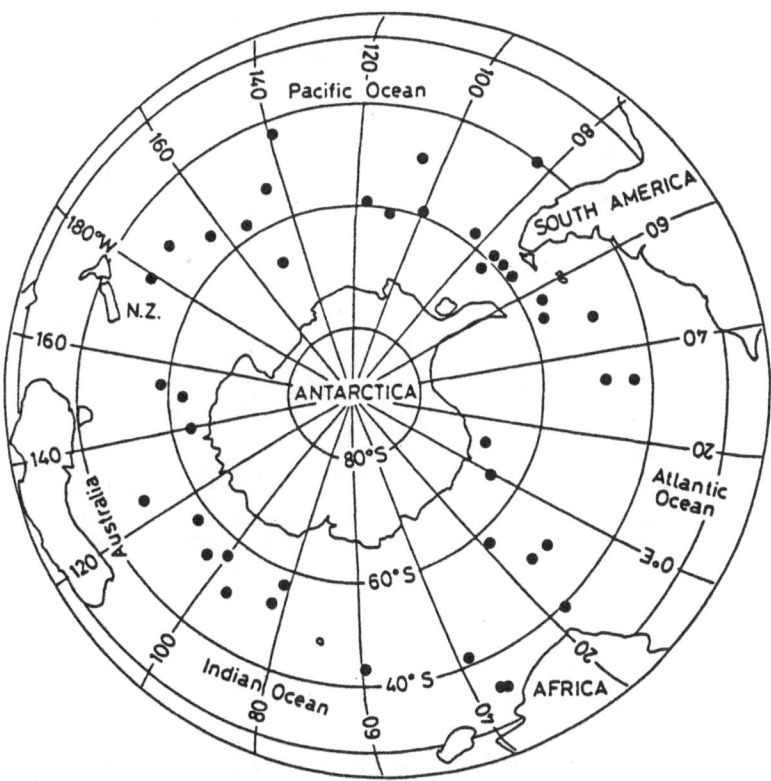

Fig. 5.32. Ferromanganese nodule sampling locations. (Abouchami and Goldstein 1995)

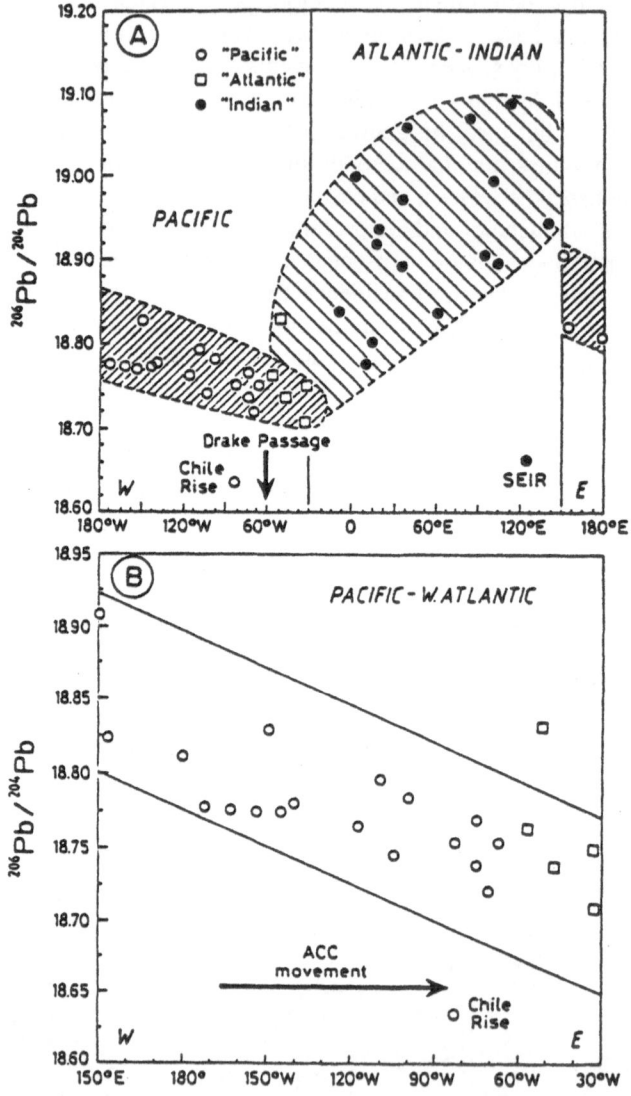

Fig. 5.33 Pb isotope ratios of marine ferromanaganese nodules show distinct variability even within one ocean basin. (Abouchami and Goldstein 1995)

Isotopic exchange between the Pacific waters and the circum-Antarctic ocean can first be ascertained in the straits between Australia and New Zealand. Manganese nodules from the western A-Atlantic and the more northerly basin of Argentina show identical Pb isotope ratios to those of the A-Pacific (Fig. 5.33, 5.34). Pb in this region comes largely from the A-Pacific and flows through the Drake passage towards the East and flows northward into the Argentinian basin.

The manganese nodules from the eastern A-Atlantic and the A-Indian show higher Pb isotopic ratios and give an indication that there is another influence here on the isotopic composition of the circumpolar ocean (Figs. 5.33; 5.34). The authors assume that the mixing of North Atlantic Deep Water (NADW) with the Circum-Antarctic waters at between 30°W und 0°, i.e. the input of Pb with higher $^{206}Pb/^{204}Pb$ ratios and lower ϵNd values helped to change the isotopic composition of circumpolar water.

Between 120°E und 150°E another abrupt switch in Pb isotopic composition can be observed. The $^{206}Pb/^{204}Pb$ ratio drops here markedly (Fig. 5.33), while Albarède und Goldstein (1992) were able to record a sharp rise in ϵNd values. It is suggested that it is in this area of the Tasman Straits between Australia and New Zealand that the Pacific seawater starts to influence the Nd and Pb isotopic composition of circum-Antarctic seawater significantly (lower part of Fig. 5.33). The study of Pb isotopes in marine authigenic minerals and especially its combination with Nd isotopes has much to offer oceanography and should be kept in mind when we turn to Ce isotopes in Sect. 5.4.

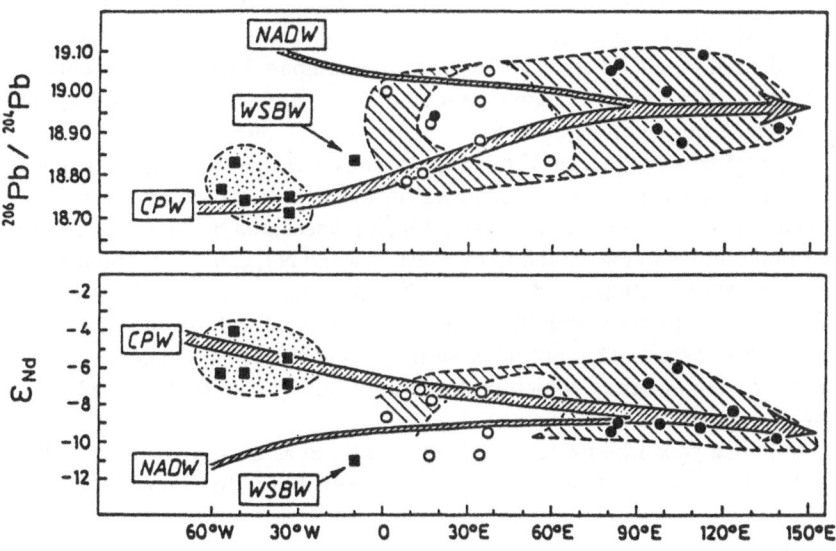

Fig. 5.34. Pb and Nd isotopic compositions of ferromanganese nodules as a function of their longitude (Abouchami and Goldstein 1995). The diagram shows how the high values of the A-Atlantic and the A-Indian waters can be explained by the mixing of the NADW (North Atlantic Deep Waters) and the CPW (Circum-Antarctic waters).

5.4 Ce Isotopic Composition of Seawater: Another Potential Tracer of Ocean Circulation?

In recent years advances on the analytic side of isotope geochemistry have meant that much isotope analysis has become standard procedure. New isotopic systems of elements whose low abundances have made their application impossible up to now are being incorporated into geologic and environmental research. Let us concentrate now on an isotopic system that is still at the very beginning of its development and application due largely to technical problems: La-Ce. Lanthanum and cerium are socalled 'rare earth elements' of slightly lower atomic mass and greater ionic radius than their sisters Sm and Nd. About one third of ^{138}La decays result in the production of ^{138}Ce. As a potential tracer of circulation patterns in the ocean for example, the Ce isotopic ratio $^{138}Ce/^{142}Ce$ has several advantages over the isotopic systems looked at so far:

1) Ce ought to have a short residence time, far less than a thousand years, as the residence time of Ce is likely to be even shorter than that of Nd. The reason for this is that it can exist in two valency states in nature, 3^+ and 4^+. All other rare earth elements (or lanthanides) exist normally as 3^+ ions except for europium (Eu), which can also exist as 2^+ ions under extremely reducing conditions and at high temperatures. In its oxidized form, cerium is prefentially removed from seawater complexed with Fe and Mn oxyhydroxide phases. Therefore, seawater shows a strong depletion in Ce (negative Ce anomaly) compared with the concentration of the other REE normalized against average shale. Because of this rapid removal, cerium could be used for fine isotopic distinction of water masses.

2) Another consequence of this rapid removal is that water masses have very different Ce concentrations depending on the age of the water mass and their oxidation state. This cerium anomaly, which is measured as the relative depletion in cerium relative to the neighbouring REE, La and Nd, can tell us about the redox conditions of deposition and diagenesis of sediments.

3) As a result, variable La/Ce ratios are typical of sedimentary authigenic or diagenetic minerals. This provides an as yet unexplored potential for dating of ancient marine minerals, which preferentially incorporate REE, e.g. phosphates, Fe-Mn oxide phases.

The La-Ce geochronometer was first introduced as a dating method in 1982 by Tanaka and Masuda. Ever since it has largely been the domain of Japanese scientists. The technical problems involved stem from the low relative abundance of ^{138}La in nature (0.09%) and its long half-life which results in a low natural range of $^{138}Ce/^{142}Ce$ ratios.

Tanaka et al. (1986) were the first to publish εCe values from ferromanganese nodules found on the seafloor today. They established that there is likely to be a difference between Ce isotopic ratios in the Pacific and Atlantic oceans and this work was continued by Amakawa et al. (1996) who more fully explored the potential of this system in marine research. These latter authors analyzed samples

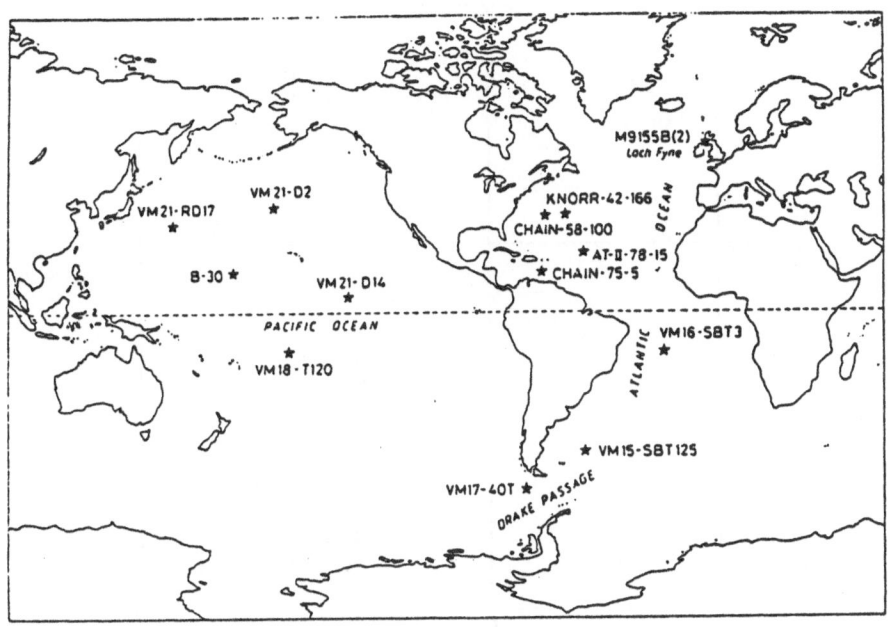

Fig. 5.35. Sampling localities. (Amakawa et al. 1996)

collected from various oceanic regions, including the Atlantic and Pacific oceans and in deep water (Fig. 5.35). They were able to confirm that a difference existed between the two oceans and were moreover able to compare these values with both the strontium and the neodymium isotopic ratios of the same samples and leachates, thus illustrating the potential of comparative isotope studies. No relationship could be found between $^{87}Sr/^{86}Sr$ ratios and εCe suggesting that diagenetic alteration did not play a major role in the isotopic development of the nodules. Their results for Nd and Ce isotopes are shown in Fig. 5.36.

As expected, samples from the Atlantic ocean show greater influence from the continent as La, being a light REE, is prefentially enriched in continental rocks. Thus, lower εNd values correlate with higher εCe values. The authors conclude however that not all sample values can be explained by the simple mixing of continental and mantle source end-members. Fractionation of Ce and Nd within the ocean environment, or for that matter in estuarine environments, needs to be considered too.

Amakawa et al. (1996) calculated a residence time of 90-165 years, which would be much shorter than those of the other REE, comparable with the residence times for ^{210}Pb and ^{231}Pa, but longer than that of Th.

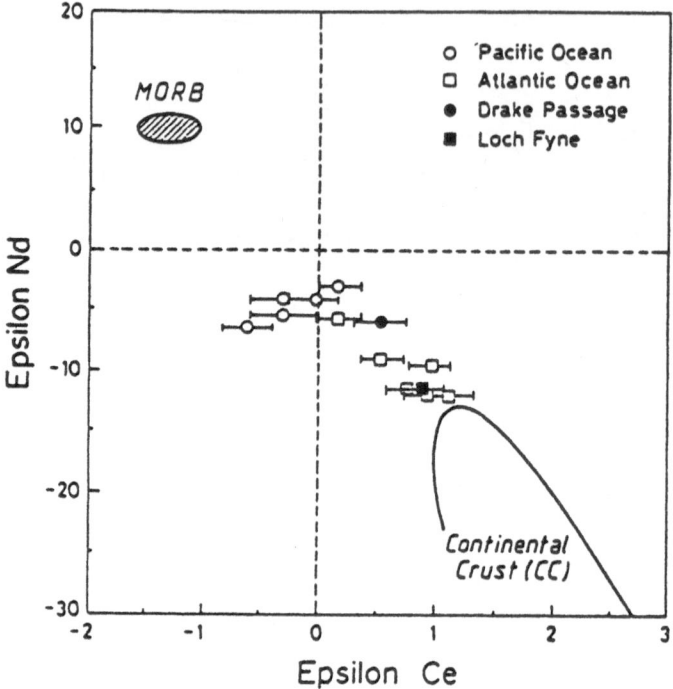

Fig. 5.36. Nd and Ce isotope results for ferromanganese nodules (Amakawa et al. 1996)

5.5 Os Isotopic Composition of Seawater: Stratigraphic Tool or Tracer of Ocean Circulation Patterns?

Rhenium (Re) has just two naturally occurring isotopes: ^{185}Re and ^{187}Re. Osmium (Os) has seven naturally occurring isotopes: ^{184}Os, ^{186}Os, ^{187}Os, ^{188}Os, ^{189}Os, ^{190}Os, and ^{192}Os. ^{187}Re undergoes beta decay with a half-life of 45.6 Ga (τ = 1.52 x 10^{-11} yr^{-1}) to form ^{187}Os in a similar fashion to ^{87}Rb and ^{87}Sr. The present day $^{187}Os/^{186}Os$ ratio depends on the age of the system, its initial $^{187}Os/^{186}Os$ ratio and its initial $^{187}Re/^{186}Os$ ratio analogous to the Sr isotope system.

Rhenium and osmium have very different compatibilities (preference for fluid or solid phase during crystallization) during magmatic processes. Rhenium is incompatible during crust-mantle differentiation processes and so becomes strongly enriched in the crust. As a result, the crust with $^{187}Re/^{186}Os$ ratios of ≈ 400, displays markedly higher $^{187}Os/^{186}Os$ ratios of 10 to 15, while the mantle shows $^{187}Os/^{186}Os$ ratios of around 1. These two sources form the two end members for the interpretation of the osmium isotopic composition in the ocean at any one time. Cosmogenic material also yields low ratios as low as the mantle. Commonly cosmogenic material is not considered in mass balances but as the concentration of osmium is relatively high in meteoritic material this ought to be considered as a third source although the solubility of this cosmogenic osmium in seawater is poorly understood.

The measurement of Os isotope ratios suffers the same problems of small concentrations in nature as many of the radiogenic isotope systems discussed so far. However, new isotope techniques involving the production of negative ions instead of positive ions has made successful analysis possible at low concentrations. Average concentrations of both Os and Re in the Earth-Moon system surface rocks are on average 1000-100,000 times less than Rb or Sr!

Concentrations of Os are likewise minute in seawater and no successful direct measurements of Os isotopic ratios have been reported in the literature. Instead, as with Pb, researchers have turned to authigenic mineral phases, which are likely to have precipitated in chemical equilibrium with seawater. For example, ferromanganese nodules may contain Os in the hundreds of ppt range. Os may also be preferentially incorporated into organic matter.

Luck and Turekian (1983) were the first to link the Os isotopic composition of ferromanganese nodules to seawater. These authors measured Os isotopic ratios from several of the World's oceans and were able to establish significant differences from place to place. As Os and Re are unlikely to be fractionated by any simple kinetic or biogenic process due to their high masses, these results could be taken to indicate that Os has a short residence time in seawater, like Nd, Pb or Ce. However, it was first necessary to demonstrate beyond doubt that the Os leachable from Fe-Mn nodules is truly representative of " hydrogenous osmium ". That is to say, osmium that derives from seawater and not from detritus or from material which has undergone exchange with detritus.

If the residence time of Os is indeed much longer than the ocean mixing time, then this system carries potential as a stratigraphic tool in the same way as Sr. Alternatively, a shorter residence time would suggest potential as a possible tracer of seawater circulation patterns cf. Nd, Pb, and Ce.

Further research on the isotopic composition of hydrogenous Os from organic-rich sediments distributed widely across the globe produced less ambigous results (Ravizza and Turekian 1992). They discovered that all samples measured, yielded an isotopic ratio of around 8.6, and so the residence time could in fact be longer than had been assumed and indeed this appears to be the case. Up to only a year ago the residence time of Os in seawater was unknown. However, recent studies have constrained it to around 10^4 years which is about an order of magnitude longer than the mixing time of the oceans (Peuker-Ehrenbrink et al. 1997).

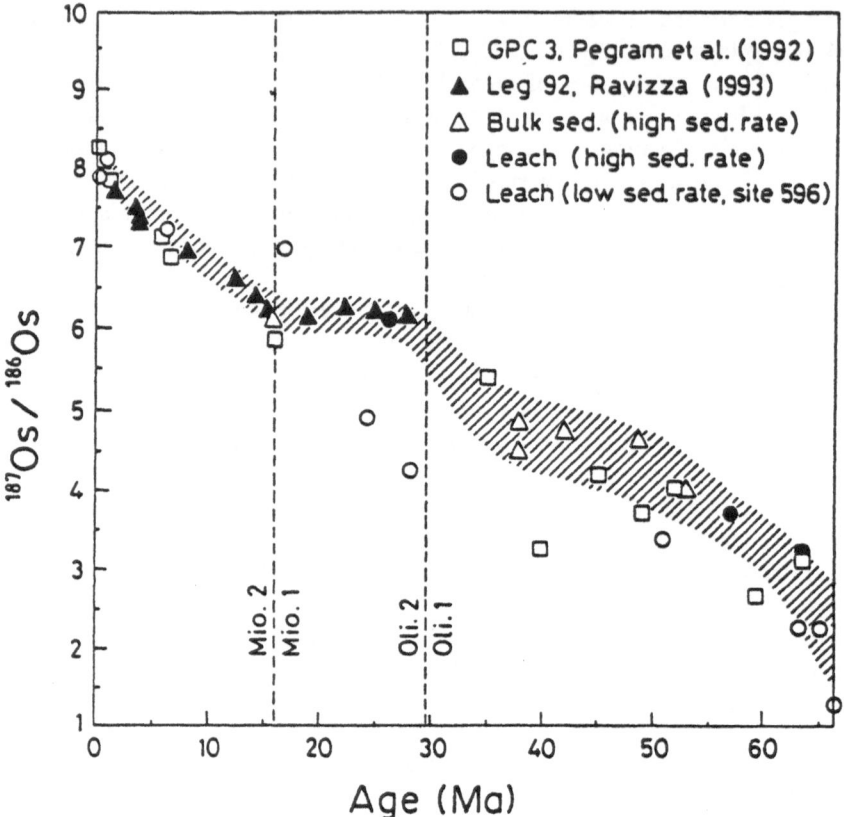

Fig. 5.37. Os isotope ratios of various authigenic mineral phases from deep ocean drill cores over the last 65 Ma. (Peuker-Ehrenbrink et al. 1995)

5.5.1 Changes in Seawater Os Isotopic Ratio Through Geologic Time

The first study that attempted to show that seawater Os isotopic ratio has varied through time was Pegram et al. (1992). The open squares of Fig. 5.37 represent the results of analyses of leachable Os from pelagic clays over nearly 60 Ma. They depict a rise in seawater $^{187}Os/^{186}Os$ from 2.8 to 8.2 through the Cenozoic era, something which has been broadly confirmed by all studies since.

Ravizza (1993) proceeded in a slightly different way. In his study he analyzed 12 bulk sediment samples from several cores drilled in the Pacific ocean during DSDP Leg 92. The sediments were all mixtures of biogenic calcium carbonate and hydrothermal Fe and Mn oxides. Such metalliferous carbonates are highly suitable for analysis as they contain predominantly hydrogenous Os. Thus no special leaching procedure is required. Nannofossils served to make stratigraphy easier with strontium isotope analysis providing an additional check on age (Sect. 5.1.1) and possible diagenetic alteration (Sect. 5.1.3).

Figure 5.38 displays the results of Sr isotope analysis of these 12 metalliferous carbonates. An HCl leach, a particularly strong leach, was applied deliberately to test the possible effect of diagenetic alteration and exchange on the samples. Only minor deviation from the Sr curve was observed and such small anomalies may have more to do with errors in age assignment. Thus, the samples pass this test. Ravizza's study leads to important considerations. First, that there was a trend to more radiogenic Os as with Sr over the last 60-30 Ma as initially shown by Pegram et al. Second, there was a change in the relationship between the two isotopic systems around 15-14 Ma (Fig. 5.38).

Earlier in this chapter we considered the origin of this rise in seawater $^{87}Sr/^{86}Sr$ and came to the conclusion that it was most likely caused by an increase in erosion and hence continental input of Sr linked to the Himalayan orogeny. Thus, it would be expected that both our isotopic proxies of continental weathering, Os and Sr, should react in tandem, but input of continental Os appears to have accelerated around 15 Ma. It is true that both systems would react to increases in weathering rates but they also react sensitively to the material being weathered. Sr is found in many rock types both sedimentary and magmatic in origin.

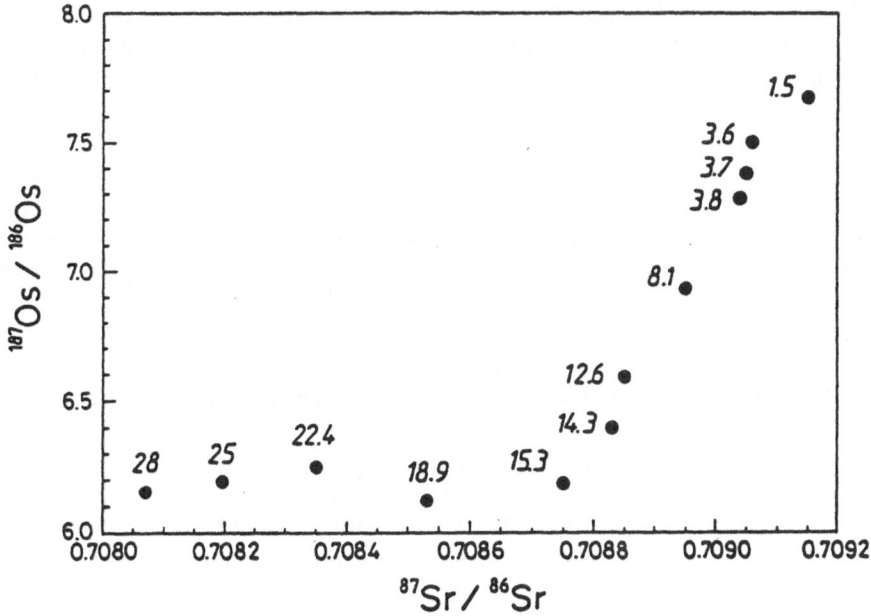

Fig. 5.38. Sr isotope ratios plotted against Os isotope ratios for twelve metalliferous carbonates from tha Pacific ocean. Numbers refer to the ages in Ma of the samples confirmed by Sr isotope analyses and biostratigraphy. (Ravizza 1993)

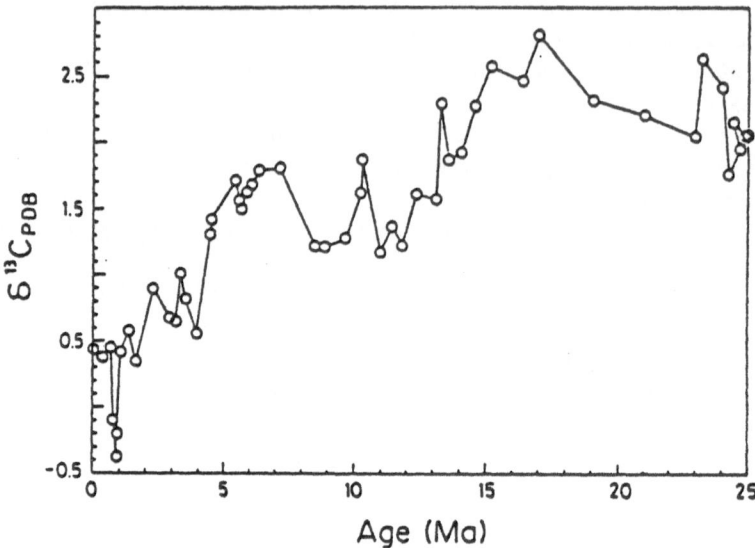

Fig. 5.39. The seawater C isotope record for the last 25 Ma based on analyses of foraminifera. (Shackleton, 1987)

However, the much rarer Os is found concentrated in few substances, most importantly, metal-rich sediments and organic matter. Fig. 5.39 shows the seawater C isotope record for the last 25 Ma (see Sect. 1.1 for a definition of $\delta^{13}C$).

Ravizza (1993) notes that there is also a change in the C isotope record at around 15 Ma. This sharp drop in $\delta^{13}C$ mirrors the increase in Os isotope ratio and can be taken to indicate a relative decrease in organic matter burial and/or an increase in the erosion of organic matter on the continent (black shales). The latter possibility would link the two changes quite neatly while explaining the more steady behaviour of the Sr isotope system.

5.5.2 Meteorite Impact at the Cretaceous-Tertiary (K-T) Boundary?

We cannot leave any discussion on the Re-Os isotopic system without exploring one of the most thought provoking issues to be raised in the earth sciences for the past twenty years: the possibility that a large extraterrestrial body caused the mass extinction of large land animals and marine microplankton around 65 Ma at the Cretaceous-Tertiary boundary.

In 1980, Alvarez et al. in a well known paper, reported that they had discovered anomalously high concentrations of iridium in sedimentary rocks deposited at K-T boundary sites in Italy, Denmark and New Zealand. Since this initial discovery, socalled iridium anomalies have been reported all over the World at this stratigraphic level, also in DSDP and ODP drill cores as well as in continental sediments. The authors' original conclusion has remained unchanged but not unchallenged. Their preferred interpretation for the high concentrations of the platinum group elements was that they had arrived from the cosmos and had been spread across the World following the impact of a meteorite, as some meteorites were known to be especially rich in these platinum group elements. Discoveries of shocked quartz, microspherules of possible tektite origin, and possibly even the smoking gun itself: a crater off Mexico, have since strengthened this hypothesis. We shall however concern ourselves mostly with the isotopic evidence for impact, bearing in mind that simultaneity of events does not prove a causal link between them.

It was believed at the time that the isotopic ratio of osmium at this level might clinch the argument. The reasons for this have been touched upon earlier in this section and relate to the extremely low, 'primitive' $^{187}Os/^{186}Os$ isotopic ratios of extraterrestrial material (due to their low initial Re/Os ratios). If the Ir at the boundary layers derived from impact then the Os isotopic ratios in the same ought to be low, around 1.

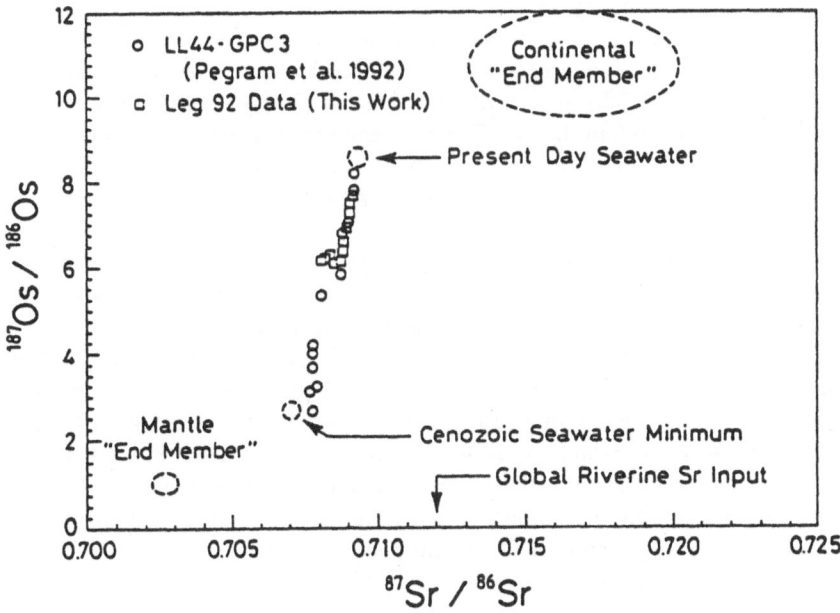

Fig. 5.40. The range of Os isotope ratios vs. Sr isotope ratio for seawater through the Cenozoic era (Ravizza 1993). It is clear that seawater Os isotope ratio has fluctuated freely between the two end members whereas Sr isotope ratio has remained relatively conservative. Is Os responding to relative rates of black shale weathering rates and cosmogenic input?

Luck and Turekian (1983) measured $^{187}Os/^{186}Os$ ratios at Stevns Klint fish clay, Denmark and Ratan Basin, Colorado yielding 1.654 +/- 0.004 and 1.29 respectively. Note these values are indeed lower than the lowest values measured by Pegram et al. (1992) or Ravizza (1993) for the succeeding 60 Ma (Fig. 5.38). Fig. 5.40 displays the range of Os isotopic ratios of mantle material relative to the continental crust. Clearly the values obtained by Luck and Turekian from the K-T boundary lie below the Cenozoic minimum but are still within a range covered by mantle material, i.e. 1-1.2. Thus, the evidence would also tend to support another theory concerning the mass extinction, that massive volcanic activity had somehow destabilized the climate and sea-level by producing huge quantities of aerosols and lavas. In this case, the Ir and Os could derive directly from the mantle. Such a large influx of mantle derived material with characteristically primitive isotopic signatures is not supported by Sr isotope evidence, which actually shows anomalously high ratios around the K/T boundary. These results can be interpreted as the result of impact related, acid rain leaching of soils and rocks with high $^{87}Sr/^{86}Sr$, typical for clays, for example.

Recently further work has been carried out at the K-T boundary by Peuker-Ehrenbrink et al. (1995). They analyzed sediments (both leachates and bulk rock) across the K-T boundary in the South Pacific where the boundary can be recognized not only on the basis of nannoplankton diversity turnover but also by the presence of shocked quartz and a pronounced Ir peak. Their work can be seen in Figs. 5.41 and 5.37. Iridium concentrations are clearly much higher at the K/T boundary at DSDP 596 in the Pacific ocean and are especially high in the zone of shocked quartz crystals. Os, both leachable and bulk, is also enriched at the boundary, whereas Os isotope ratios mirror these trends, being lowermost in the sample richest in osmium and iridium.

It may also be noticed that low Os isotope ratios preceded the actual boundary. If these ratios were truly representative of seawater, it would certainly lessen the credibility of any impact signature. The authors argue convincingly that this smearing of the signal may be the result of bioturbation, something which is additionally supported by the presence of shocked quartz up to 1 m below the K-T boundary. The low Os isotope ratios in samples whose concentrations are very high at the K-T boundary is perfectly consistent with an impact scenario. Estimates range around $5 * 10^{11}g$ for the amount of cosmogenic Os that might have been released after the impact which would correspond to the fluviatile input of 6 Ma. Such a huge input of Os would necessarily have had an enormous and immediate effect on the Os isotopic system. Such large amounts would explain the sluggish recovery of the seawater Os isotope ratio after the boundary, but only provided the residence time of Os is long (over 1 Ma), although this does not appear to be the case with estimates on the order of 10^4 years (Peuker-Ehrenbrink et al. 1997) and only if low Os isotope ratios below the boundary are an artefact of bioturbation. More work clearly needs to be carried out to separate cosmogenic Os from hydrogenous and detrital Os for isotopic measurement but the Re-Os and Ir evidence still remains at the heart of the K-T impact hypothesis.

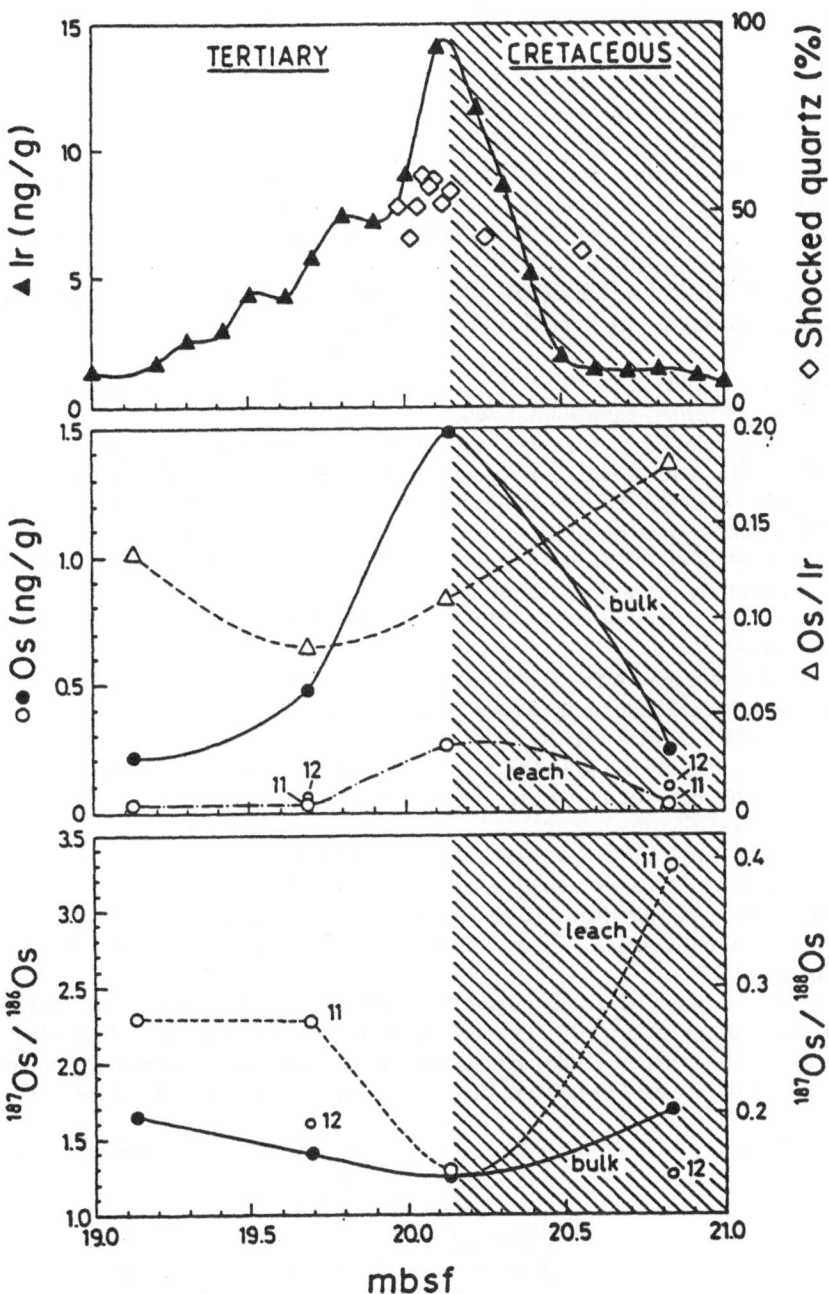

Fig. 5.41. Detailed isotopic study of the K-T boundary of DSDP site 596 in the Pacific ocean. Os isotope ratios are lowest and Os and Ir concentrations are highest in the zone of shocked quartz marking the boundary. (Peuker-Ehrenbrink et al. 1995)

5.6 References

Aberg F, Wickman FE (1987) Variations of $^{87}Sr/^{86}Sr$ in water from streams discharging into the Botham Bay, Baltic Sea. Nordic Hydrol., 18: 33-42

Abouchami W, Goldstein SL (1995) A lead isotopic study of Circum-Antarctic manganese nodules. Geochim Cosmochim Acta, 59: 1809-1820

Albarède F, Goldstein SL (1992) A world map of Nd isotopes in seafloor ferromanganese deposits. Geology, 20: 761-763

Albarède F, Michard A, Minster JF, Michard G (1981) $^{87}Sr/^{86}Sr$ ratios in hydrothermal waters and deposits from the East Pacific Rise at 21°N. Earth Planet Sci Lett, 55: 229-236

Alvarez LW, Alvarez W, Asaro F, Michel HV (1980) Extraterrestrial cause for the Cretaceous-Tertiary extinctions. Science 208: 1095-1108

Amakawa H, Nozaki Y, Masuda, A (1996) Precise determination of variations in the $^{138}Ce/^{142}Ce$ ratios of marine ferromanganese nodules. Chem Geol 131: 183-195

Arthur MA, Anderson TF (eds)(1983) Stable isotopes in sedimentary geology. Soc Econ Petrog Min Short course no. 10

Banner JL, Hanson GN, Meyers WJ (1988) Rare earth element and Nd isotopic variations in regionally extensive dolomites from the Burlington-Keokuk formation (Mississippian): implications for REE mobility during carbonate diagenesis. J Sed Petrol, 58: 415-432

Brasier MD, Rozanov AYu, Zhuravlev AYu, Corfield RM, Derry LA (1994) A carbon isotope reference scale for the Lower Cambrian succession in Siberia: report of the IGCP Project 303. Geol Mag 131: 767-783

Brasier MD, Shields GA, Kuleshov VN, Zhegallo EA (1996) Integrated chemo- and bio-stratigraphic calibration of early animal evolution of southwest Mongolia. Geol Mag 133: 445-485

Broeker WS, Peng TS (1982) Tracers in the sea; Eldigo Press

Bros R (1993) Géochimie isotopique (Sr, Nd, Ar, Pb, U) appliqué a des processus diagenetiques et hydrothermaux. Evolution du Bassin de Franceville d'age proterozoique inférieur et in fluences des réacteurs nucleaires fossiles d'Oklo (Gabon). PhD Thesis, Uni Strasbourg

Burke WH, Denison RE, Hetherington EA, Koepnick RB, Nelson NF, Otto JB (1982) Variation of seawater $^{87}Sr/^{86}Sr$ throughout Phanerozoic time. Geology, 10: 516-519

Chaudhuri S, Clauer N (1986) Fluctuations of isotopic composition of strontium in seawater during the Phanerozoic Eon. Chem Geol (Isotope Geoscience Section), 59: 293-303

Chow TJ, Patterson CC (1962) The occurrence and significance of lead isotopes in pelagic sediments. Geochim Cosmochim Acta, 26: 263-308

Chyi MS, Crerar DA, Carlson RW, Stallard RF (1984) Hydrothermal Mn-deposits of the Franciscan Assemblage, II. Isotope and trace element geochemistry, and implications for hydrothermal convection at spreading centers. Earth Planet Sci Lett, 71: 31-45

Clauer N, Stille P, Bonnot-Courtois C, Moore WS (1984) Nd-Sr isotopic and REE constraints on the genesis of hydrothermal manganese crusts in the Galapagos. Nature, 311: 743-745

Craig H, Krishnaswami S, Somayajulu BLK (1973) ^{210}Pb-^{226}Ra: Radioactive disequilibrium in the deep sea. Earth Planet Sci Lett, 17: 295-305

DePaolo DJ, Wasserburg GJ (1977) The sources of island arcs as indicated by Nd and Sr isotopic studies. Geophys Res Lett, 4: 465-468

DePaolo DJ, Ingram BL (1985) High resolution stratigraphy with strontium isotopes. Science, 227: 938-941

Derry LA, Kaufman AJ, Jacobsen SB (1992) Sedimentary cycling and environmental change in the Late Proterozoic: evidence from stable and radiogenic isotopes. Geochim Cosmochim Acta 56: 1317-1329

Dia AN, Cohen AS, O'Nions RK, Shackleton NJ (1992) Seawater Sr isotope variation over the past 300 Kyr and influence of global climate. Nature 356: 786-788

Elderfield H, Greaves MJ (1981) Strontium isotope geochemistry of Icelandic geothermal systems and implications for seawater chemistry. Geochim Cosmochim Acta, 45: 2201-2212

Faure G (1982) The marine-strontium geochronometer. In: GS Odin, ed., Numerical Dating in Stratigraphy, John Wiley & Sons, New York, 1: 73-79

Faure G, Hurley PM, Powell JL (1965) The isotopic composition of strontium in surface water from the North Atlantic Ocean. Geochim Cosmochim Acta, 29: 209-220

Fiechtner L, Friedrichsen H, Hammerschmidt K (1992) Geochemistry and geochronology of Early Mesozoic tholeiites from Central Morocco. Geol Rundsch, 81: 45-62

Flegal AR, Itoh K, Patterson CC, Wongs CS (1986) Vertical profile of lead isotopic compositions in the North-East Pacific. Nature, 321: 689-690

Flegal AR, Nriagu JO, Niemeyer S, Coale KH (1989) Isotopic tracers of lead contamination in the Great Lakes. Nature, 339: 455-458

Froelich PN, Klinkhammer GP, Bender ML, Luedtke NA, Heath GR, Cullen D, Dauphin P, Hammond D, Hartman B, Maynard V (1979) Early oxidation of organic matter in pelagic sediments of the eastern equatorial Atlantic: suboxic diagenesis. Geochim Cosmochim Acta, 43: 1075-1090

Frost CD, O'Nions RK, Goldstein SL (1986) Mass balance for Nd in the Mediterranean Sea. Chem Geol, 55: 45-50

Goldstein SL, O'Nions RK (1981) Nd and Sr isotopic relationships in pelagic clays and ferromanganese deposits, Nature, 292: 324-327

Gorokhov IM, Clauer N, Turchenko TL, Melnikov NN, Kutyavin EP, Pirrus E, Baskakov AV (1994) Rb-Sr systematics of Vendian-Cambrian claystones from the east European Platform: implications for a multi-stage illite evolution. Chem Geol, 112: 71-89

Grandjean P, Cappetta H, Michard A, Albarede F (1987) The assessment of REE patterns and $^{143}Nd/^{144}Nd$ ratios in fish remains. Earth Planet Sci Lett, 84: 181-196

Henderson GM, Martel DJ, O'Nions RK, Shackleton NJ (1994) Evolution of seawater $^{87}Sr/^{86}Sr$ over the last 400 ka: the absence of glacial/interglacial cycles. Earth Planet Sci Lett 128: 643-651

Hodell DA, Mueller PA, Garrido JR (1991) Variations in the strontium isotopic composition of seawater during the Neogene. Geology, 19: 24-27

Hoefs J (1980) Stable isotope geochemistry 2nd edition. Springer-Verlag, 208 pp

Hooker PJ, Hamilton PJ, O'Nions RK (1981) An estimate of the Nd isotopic composition of Iapetus seawater from ca. 490 Ma metalliferous sediments. Earth Planet Sci Lett, 56: 180-188

Jacobs E, Weissert H, Shields G, Stille P (1996) The Monterey event in the Mediterranean: a record from shallow shelf sediments from Malta. Palaeocean,11/6:717-728

Jacobsen SB, Pimentel-Klose MR (1988a) Nd isotopic variations in Precambrian Banded Iron Formations. Geophys Res Lett, 15: 393-396

Jacobsen SB, Pimentel-Klose MR (1988b) A Nd isotopic study of the Hamersley and Michipicoten banded iron formations: the source of REE and Fe in Archean oceans. Earth Planet Sci Lett, 87: 29-44.

Keto LS, Jacobsen SB (1987) Nd and Sr isotopic variations of early Paleozoic oceans. Earth Planet Sci Lett 84: 27-41

Keto LS, Jacobsen SB (1988) Nd isotopic variations of Phanerozoic paleoceans. Earth Planet Sci Lett, 90: 395- 410

Luck JM, Turekian KK (1983) Osmium-187/osmium-186 in manganese nodules and the Cretaceous-Tertiary boundary. Science 222: 613-615

McArthur JM, Sahami AR, Thirwall M, Hamilton PJ, Osborn AO (1990) Dating phosphogenesis with Sr isotopes. Geochim Cosmochim Acta, 54: 1343-1351

McArthur JM (1994) Recent trends in strontium isotope stratigraphy. Terra Nova 6: 331-358

Miller RG, O'Nions RK (1985) Source of Precambrian chemical and clastic sediments. Nature, 314: 325-330

Nicholas CJ (1996) The Sr isotopic evolution of the oceans during the 'Cambrian explosion'. J Geol Soc London 153: 243-254

O'Nions RK, Carter SR, Cohen RS, Evensen NM, Hamilton PJ (1978) Pb, Nd and Sr isotopes in oceanic Fe-Mn deposits and ocean floor basalts. Nature, 273: 435-438

Palmer MR, Elderfield H (1986) Rare earth elements and neodymium isotopes in ferromanganese oxide coatings of Cenozoic foraminifera from the Atlantic Ocean. Geochim Cosmochim Acta, 50: 409-417

Pegram WJ, Krishnaswami S, Ravizza G (1992) The record of seawater $^{187}Os/^{186}Os$ variatioàn through the Cenozoic. Earth Planet Sci Lett 113: 569-576

Peterman ZE, Hedge CE, Tourtelot HA (1970) Isotopic composition of strontium in seawater throughout Phanerozoic time. Geochim Cosmochim Acta, 34: 105-120

Peuker-Ehrenbrink B, Ravizza G, Hofmann AW (1995) The marine $^{187}Os/^{186}Os$ record of the past 80 million years. Earth Planet Sci Lett, 130: 155-167

Peuker-Ehrenbrink B, Blum JD, Bollhöfer A (1997) The effects of global glaciations on the marine Os isotope record. Terra Abstracts, 9: 616

Piepgras DJ, Wasserburg GJ, Dasch EJ (1979) The isotopic composition of Nd in different ocean masses. Earth Planet Sci Lett, 45: 223-236

Piepgras DJ, Wasserburg, GJ (1980) Nd isotopic variations in seawater. Earth Planet Sci Lett, 50: 128-138

Piepgras DJ, Wasserburg GJ (1982) Isotopic composition of neodymium in waters from the Drake Passage. Science, 217: 207-214

Piepgras DJ, Wasserburg GJ (1983) Influence of the Mediterranean outflow on the isotopic composition of Nd in waters of the North Atlantic. J Geophys Res, 88: 5997-6006

Piper JDA (1982) The Precambrian palaeomagnetic record: the case of the Proterozoic Supercontinent. Earth Planet Sci Lett, 59: 61-89

Ravizza G (1993) Variations of the $^{187}Os/^{186}Os$ ratio of seawater over the past 28 million years as inferred from metalliferous carbonates. Earth Planet Sci Lett, 118: 335-348

Riggs SR (1984) Paleoceanographic model of Neogene phosphorite deposition, U.S. Atlantic Continental Margin. Science, 223: 123-131

Riggs SR, Stille P, Ames D (1997) Sr isotopic age analysis of co-occurring Miocene phosphate grain types on the North Carolina Continental Shelf. J Sed Res, 67: 65-73

Shackleton N (1987) The carbon isotope record of the Cenozoic: History of organic carbon burial and oxygen in the ocean and atmosphere, in: Marine Petroleum Source Rocks, J. Brooks and A.J. Fleet (eds.), pp. 423-434

Shaw HF, Wasserburg GJ (1985) Sm-Nd in marine carbonates and phosphates: Implications for Nd isotopes in seawater and crustal ages. Geochim Cosmochim Acta, 49: 503-518

Shields GA (1996) Event stratigraphy around the Precambrian/Cambrian boundary with special emphasis on South China and western Mongolia. PhD thesis, ETH Zürich

Staudigel H, Doyle P, Zindler A (1986) Sr and Nd isotope systematics in fish teeth. Earth Planet Sci Lett, 76: 45-56

Stille P, Clauer N (1986) Sm-Nd isochron age and provenance of the argillites of the Gunflint Iron Formation in Ontario, Canada. Geochim Cosmochim Acta, 50: 1141-1146

Stille, P (1987) Geochemische Aspekte der Krustenevolution im zentral- und südalpinen Raum. Habilitationsschrift, ETH Zürich, pp. 127

Stille P, Clauer N, Abrecht J (1989) Nd isotopic composition of Jurassic Tethys seawater and the genesis of Alpine Mn deposits: evidence from Sr-Nd isotope data. Geochim Cosmochim Acta, 53: 1095-1099

Stille P, Fischer H (1990) Secular variation in the isotopic composition of Nd in Tethys seawater. Geochim Cosmochim Acta, 54: 3139-3145

Stille P (1992) Nd-Sr isotope evidence for dramatic changes of paleo-currents in the Atlantic Ocean during the past 80 million years. Geology, 20: 387-390

Stille P, Chaudhuri S, Kharaka YK, Clauer, N (1992) Isotope compositions of waters in present and past oceans: A review. In: "Isotopic signatures and sedimentary records". (Eds. Clauer N and Chaudhuri S) Springer Verlag

Stille P, Riggs S, Clauer N, Crowson R, Snyder SW, Ames D (1994) Sr and Nd isotopic analysis of phosphorite sedimentation through one Miocene high-frequency depositional cycle on the North Carolina Continental Shelf. Marine Geology, 117: 253-273

Stille P, Steinmann M, Riggs SR (1996) Nd isotope evidence for the evolution of the paleocurrents in the Atlantic and Tethys oceans during the past 180 Ma. Earth Planet Sci Lett. 144: 9-19

Tanaka T, Masuda, A (1982) The La-Ce geochronometer: a new dating method. Nature 300: 515-518

Tanaka T, Usui A, Masuda A (1986) Oceanic Ce and continental Nd: Multiple sources of REE in oceani ferromanganese nodules. Terra Cognita abstracts 6: 114

Vail PR, Mitchum RM, Todd RG, Widmier JM, Thompson S, Sangree JB, Bubb J, Hatlelid WG (1977) Seismic stratigraphy and global changes of sea level. Am Assoc Pet Geol Mem, 26: 49-212

Veizer J, Compston, W (1974) $^{87}Sr/^{86}Sr$ composition of seawater during the Phanerozoic. Geochim Cosmochim Acta, 38: 1461-1484

Veizer J, Compston W (1976) $^{87}Sr/^{86}Sr$ in Precambrian carbonates as an index of crustal evolution. Geochim Cosmochim Acta, 4: 905-914

Veizer J (1983) Chemical diagenesis of carbonates: theory and application of trace element technique. In Stable isotopes in sedimentary geology Arthur MA, Anderson TF (eds) Soc econ petrog min, Short course no. 10, 3: 1-100

Veizer J, Compston W, Clauer N, Schidlowski M (1983) $^{87}Sr/^{86}Sr$ in Late Proterozoic carbonates: evidence for a "mantle" event at ~ 900 Ma ago. Geochim Cosmochim Acta, 47: 295-302

6 Isotope Geochemistry of Detrital and Authigenic Clay Minerals in Marine Sediments (Rb-Sr, K-Ar, O)

It was recognized early on in modern, marine research that most Holocene marine sediments are of detrital origin (e.g. Biscaye 1965). These detrital sediments are commonly very fine grained and are derived from the weathering of rocks on the continents. Biscaye (1965) suggested that the composition of the <2 μm clay fraction of this detritus may reflect the conditions of weathering on the continent and so can be used to detect possible climatic fluctuations. The mineralogy of detrital clay minerals can be used to determine both their place and nature of origin as they give clues as to both the precursor rock and the mode and intensity of weathering. The application of isotope methods to these detrital clay fractions and a comparison of their isotopic ratios with those of river, ocean and pore waters ought to make it possible not only to determine the origin of the clay minerals, but also to help us appreciate the chemical exchange processes between clay minerals and their surroundings better.

6.1 Detrital Clay Minerals

The isotope study of Dasch (1969) on the detrital silicate fraction of marine sediments from the Atlantic represents a pioneering piece of work in the field of low temperature isotope geochemistry. This study is still used today to prove that clay minerals rarely form in isotopic equilibrium with their sedimentary environment and so can not be used for dating sedimentary rocks (Savin and Lee 1988).

In actual fact, Dasch investigated the whole silicate fraction of the sediments after treatment with HCl and removal of the soluble fraction. He determined therefore the whole detrital fraction, which explains the large variations in $^{87}Sr/^{86}Sr$ found in the investigated sediments. Dasch carried out isotopic analyses

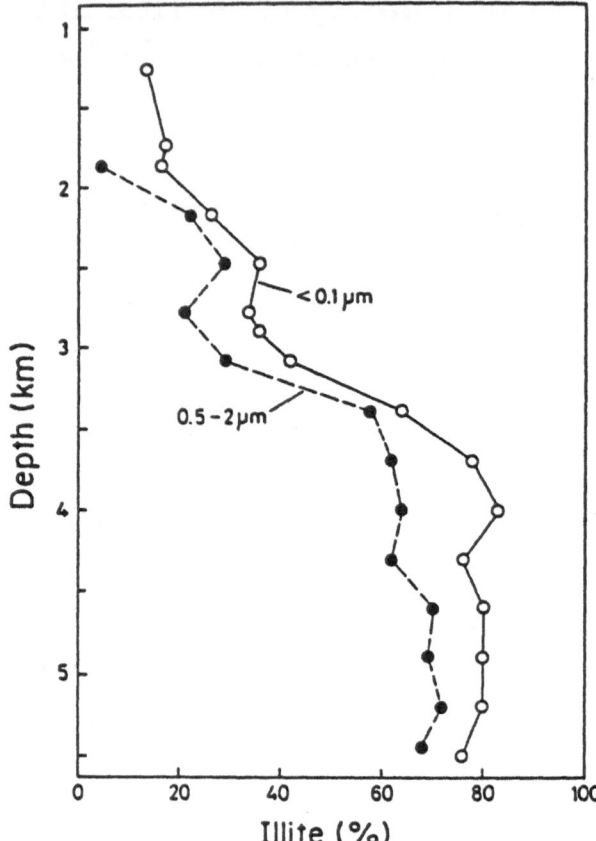

Fig. 6.1. Depth dependence of the proportion of illite layers in illite/smectite. (Hower et al. 1976)

on material which contained not only detrital clay minerals but also other silicate minerals such as micas and feldspars with very high $^{87}Sr/^{86}Sr$ ratios.

Many of the detrital weathering products formed at the Earth's surface under the influence of isotopically light groundwater (low ^{18}O content; see Sect. 2.4). These detrital micas should contain far lower $\delta^{18}O$ values than those of authigenic clay minerals which would have formed in isotopic equilibrium with the isotopically heavier seawater (higher ^{18}O content). The investigations of Savin and Epstein (1970) demonstrated that detrital micas possessed $\delta^{18}O$ values of 16-18‰ (SMOW) while authigenic clay minerals showed much higher values of 26-31‰. The oxygen isotope method can therefore be used as an important tool in sedimentary research, especially for the differentiation of detrital and authigenic clay minerals.

In the isotope study of Yeh and Savin (1976) and Yeh and Epstein (1978), carried out on Quaternary, detrital clay minerals from marine sediments, it was only possible to recognize minor oxygen and hydrogen isotope exchange with seawater in particles smaller than 0.1 μm in diameter. Yeh and Eslinger (1986) carried out similar investigations on the <0.3 μm clay mineral fraction of deep sea sediments. The measured $\delta^{18}O$ values scatter around 20‰ (SMOW) and do not lie on the theoretical curve, which can be calculated for oxygen isotopic equilibrium between clay minerals, porewater and seawater. Even the sample from the deepest borehole shows low $\delta^{18}O$ values, where one might expect some diagenetic recrystallization and isotopic exchange. Such low values can easily be differentiated from the higher ones of authigenic clay minerals.

These data indicate that: 1) the isotopic composition of the clay minerals has not been significantly altered, and 2) the detrital minerals have not been overgrown by authigenic clay minerals. No significant isotopic exchange could have taken place between clay minerals and seawater throughout this 600 meter deep bore hole. The <0.3 μm fraction has thus far been able to retain its purely detrital isotopic signature for 3 million years.

Clay minerals behave very differently at even greater burial depths under the influence of diagenesis. This was the case in several mineralogic, geochemical and isotopic studies of pelitic sediments in bore holes, sunk into the Mississippi delta of the US Gulf coast of Texas.

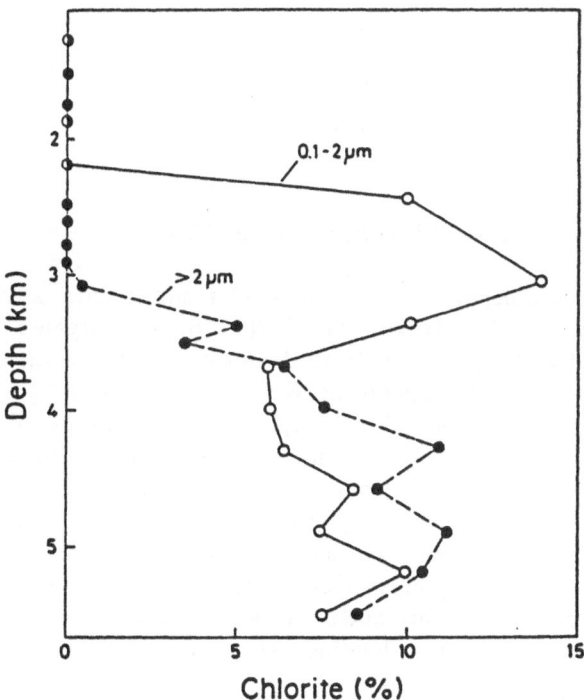

Fig. 6.2. Depth dependence of chlorite contents. (Hower et al. 1976)

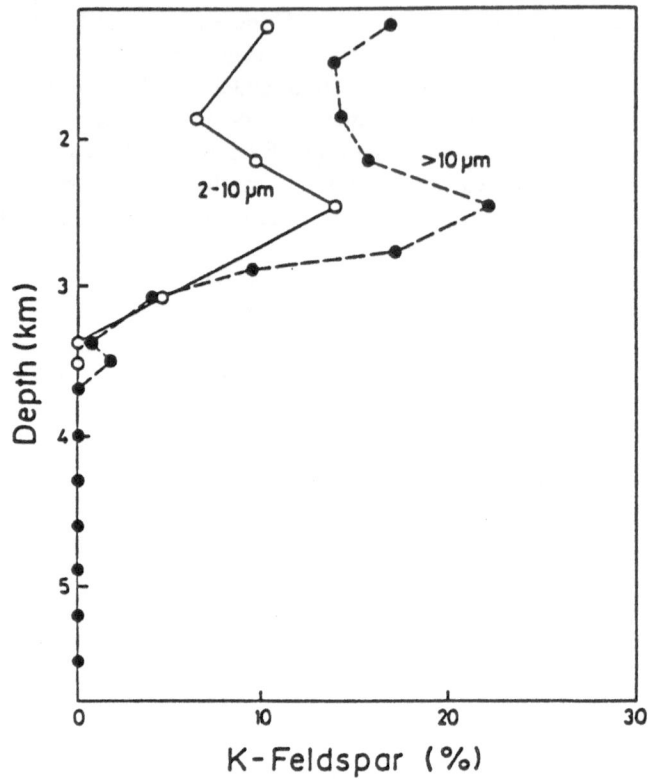

Fig. 6.3. Depth dependence of K-feldspar contents. (Hower et al. 1976)

These cores reach depths of more than 5000 m and are cited as classic examples of burial diagenesis. Hower et al. (1976) investigated these profiles mineralogically. They were able to demonstrate changes in the mineralogic composition of the lithified sediments with depth. Most striking is the mineralogic change observed at around 3000 m depth. At this depth, the illite content rises significantly at the expense of smectite (Fig. 6.1). Likewise, chlorite appears (Fig. 6.2), while alkali feldspar content drops below the detection limit (Fig. 6.3). The calcite content also falls markedly at this depth (Fig. 6.4), while the amount of quartz remains more or less constant through the whole profile (Fig. 6.5). The authors present various pieces of evidence, which show that these changes in mineralogic composition are not the result of variations in the chemistry and mineralogy of the detritus, but instead are the consequences of increasing diagenetic alteration with depth.

The depth dependence of the K_2O content of various clay fractions is displayed in Fig. 6.6. Although the overall potassium content remains more or less constant along the whole profile, a rise in the potassium content in the finest fraction can be observed at about 3000 m depth, while the K content of the coarser fraction falls at the same depth. The increase in K content with depth in the finest fraction corresponds to an increase in illite, while the simultaneous loss of K from the coarse fraction corresponds to the progressive disappearance of alkali feldspar. Note that alkali feldspars are found mainly in the coarse fraction while the finest fraction contains almost exclusively illite and smectite.

Hower et al. assume that the following reaction takes place with increasing depth and diagenetic alteration:

$$\text{Smectite} + Al^{3+} + K^+ = \text{Illite} + Si^{4+} + H_2O \qquad (I)$$

Enrichment in K and Al in the finest fraction accompanies therefore the transformation of smectite into illite (Fig. 7.1). Potassium and aluminium are required for the formation of illite and derive from the alkali feldspar and muscovite, which react with and are decomposed by organic acids, which have been released at great depths by the decarboxylation of organic matter. The same acidic solutions are also responsible for the disappearance of carbonate at great depths (Fig. 7.4).

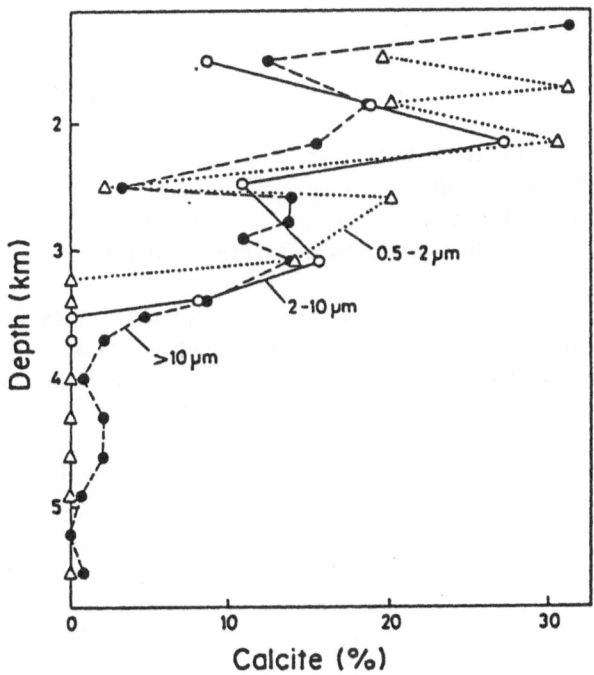

Fig. 6.4. Depth dependence of calcite contents. (Hower et al. 1976)

The following reactions are likely to take place:

$$CaCO_3 + H_2O + CO_2 \text{-------}> Ca^{++} + 2HCO_3^{-} \qquad \text{(II)}$$

$$2KAlSi_3O_8 \text{ (K-feldspar)} + 2CO_2 + 11H_2O \text{-----}>$$
$$Al_2Si_2O_5(OH)_4 \text{ (smectite)} + 2K^{+} + 4H_4SiO_4 + 2HCO_3^{-} \text{ (III)}$$

Thus, the decomposition of alkali feldspar does not only lead to the release of potassium but also to the further formation of smectite. The CO_2 which is necessary for these reactions to take place may be provided by the decomposition, oxidation and maturation of organic carbon (CH_2O) at great depths:

1) by the formation of soluble iron through the reduction of iron (III) oxides:

$$2 CH_2O + 8 FeOOH + 16 H^{+} = 2 CO_2 + 14 H_2O + 8Fe^{2+} \qquad \text{(IV)}$$

2) by the reduction of sulphate and formation of hydrogen sulphide:

$$2 CH_2O + SO_4^{2-} + H^{+} = HS^{-} + 2 CO_2 + 2 H_2O \quad \text{(V)}$$

Protons which are necessary for these reactions to take place can be produced by the formation of alumino-silicates (e.g. chlorite) as the following equation shows:

$$SiO_2 + clay + cations \text{ ---}> Al\text{-silicates} + water + H^{+} \qquad \text{(VI)}$$

Fig. 6.2. confirms this reaction path and shows that the chlorite formation stage sets in at just 3000 m depth. The disappearance of K-feldspar and the increasing amount of illite at the expense of smectite can be explained by diagenetic processes which according to Hower et al. (1976) took place in a nearly closed diagenetic system.

Yeh and Savin (1977) carried out oxygen isotope analyses on clay minerals of various grain sizes. The results of these experiments can be seen displayed in Fig. 6.7 as a function of temperature and depth. The large variation in $\delta^{18}O$ values in the upper part of the bore profile points to the detrital origin of the micas there. The isotopic ratios imply that there were various sources for the detrital material and allow us to disregard the effect of isotopic exchange with pore waters. The micas begin to show signs of exchange with pore water with increasing temperature and depth. At temperatures above 85° C and depths of burial greater than 3000 m the region of variation in isotopic ratios decreases markedly and the isotopic signatures of the various grains start to merge. Yeh and Savin (1977) assume this means that significant isotopic exchange has indeed taken place. This direct isotopic exchange can be related to the dehydration of the clays and crystallographic alterations in the clay minerals, namely the transformation of smectite into illite at temperatures of around 80° C. Hydrogen behaves similarly

(Fig. 6.8; Yeh 1980). A major switch in hydrogen isotopic composition occurs likewise at about 3000 m depth. The changes in the hydrogen and oxygen isotope compositions are likely to be related to the transformation of smectite into illite. Once again we see the confirmation of a very general rule of the oxygen-hydrogen isotopic system, i.e. that changes in isotopic composition must be related to some kind of mineralogic transformation or restructuring of a mineral lattice.

With the help of oxygen isotopic analyses on various quartz grain-size fractions, Yeh and Lavin (1977) could demonstrate that new quartz growth took place at the same time as the smectite-illite transition (see equations I and III). This study proves interesting in another regard, too. The authors observed that the $\delta^{18}O$ values, calculated for pore water in equilibrium with the <0.1 μm fractions of illite and smectite, rise consistently with increasing depth and temperature. The oxygen isotopic exchange between smectite, illite and pore water must have occurred in an almost completely closed diagenetic system. On the basis of good correlation between the $\delta^{18}O$ of the pore water, the temperature and the bore profile depth, it can be assumed that vertical pore water circulation must have been limited in the clay-rich sedimentary sequence, at least during the period of illitization.

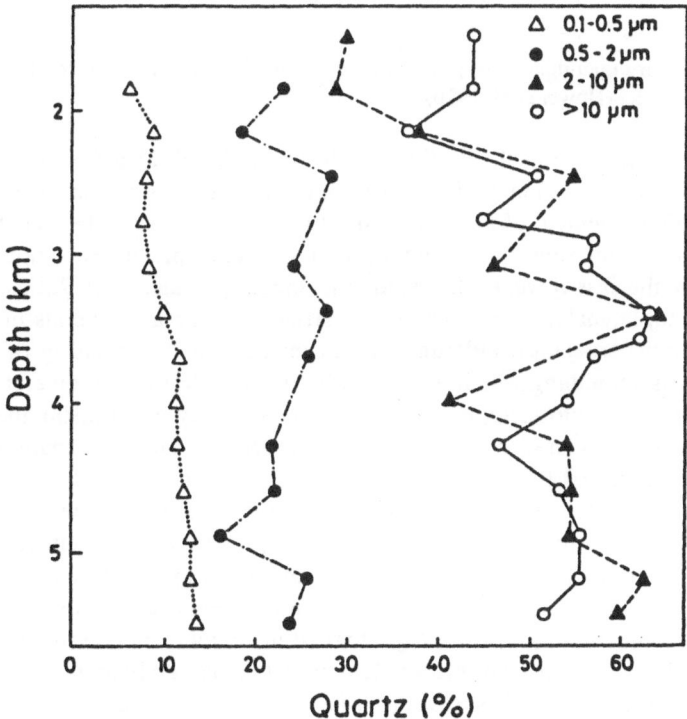

Fig. 6.5. Quartz contents of various size fractions as a function of depth. (Hower et al. 1976)

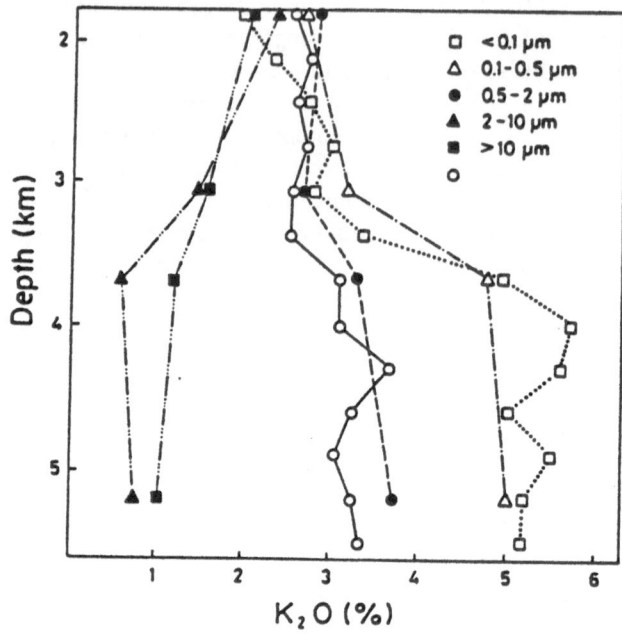

Fig. 6.6. Depth-dependence of potassium contents in various size fractions. Open circle:whole shale. (Hower et al. 1976)

This observation confirms the findings of Hower et al. (1976) and is in this respect of great importance, as many clay mineralogists assume that a significant amount of large-scale movement of pore water must occur in order that the smectite-illite transformation and subsequent dehydration can take place. This appears not always to be the case if we use this study as a model (see also Sect. 7.2).

K-Ar isotopic analyses are commonly carried out on clay minerals in order to date diagenetic processes, although the mechanism of the resetting of isotopic equilibrium is something which is frequently little constrained or understood. It is important to take a closer look at the classic K-Ar study of Aronson and Hower (1976). The K concentrations, radiogenic argon contents and the apparent ages of the whole rock and <0.1 μm clay fraction are displayed in Table 6.1.

The apparent ages are mixing ages and result from both inherited, ancient detritus (>150 Ma) and young mineral formation, whose ages are either the same or younger than the age of sedimentation (stratigraphic age). These apparent ages decrease with increasing depth and are made ever younger by the already discussed diagenetic processes. No geological meaning can be assigned to these ages. The falling K-Ar ages of the whole rock with depth go hand in hand with the loss of radiogenic argon (Figs. 6.9 and 6.10). This loss can scarcely be attributed to degassing as a result of greater temperatures as the finest fraction, which consists predominantly of illite and smectite, shows an increase in radiogenic argon with increasing depth. It seems more likely that this argon loss takes place in the coarser fraction, which is particularly enriched in detrital alkali feldspar and micas. These detrital mineral phases display high contents of radiogenic argon. The destruction and dissolution of these mineral phases must have released argon. If we look at the clay fractions a little more closely: the rise in K content with depth has the consequence of a rise in radiogenic argon (through the radioactive decay of ^{40}K to argon) and so leads to a fall in the apparent age, which reaches 30 Ma at depths around 5000 m. Under the assumption that all radiogenic argon in the newly formed illites has been produced by the radioactive decay of potassium in the illite structure (ΔAr at depths of between 3000 and 4000 meters; Fig. 6.11), an average age of 18.4 +/-2 Ma can be estimated for diagenesis and smectite-illite transformation. This diagenesis caused both an increase in potassium (ΔK_2O in Fig. 6.11) and radiogenic argon (Δ^{40}Ar in Fig. 6.10) contents in the fine fractions. On the basis of sedimentary observations it appears that this age could be geologically relevant. The investigated sedimentary basin was still undergoing subsidence and sedimentation about 25 Ma ago. The last 2150 meters of the sedimentary pile were deposited between 25 and 15 Ma ago, which means that the sedimentary series, which today lies between depths of 2150 and 5550 m, were at the surface and at 3400 m depth, respectively, around 25 Ma ago. The diagenetic process must therefore have set in less than 25 Ma ago.

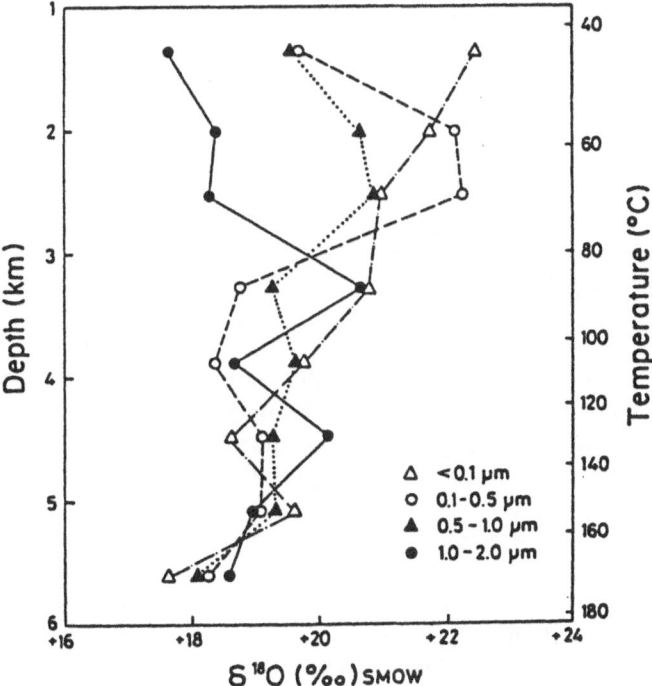

Fig. 6.7. Depth dependence of $\delta^{18}O$ in different clay size-fractions. (Yeh and Savin 1977)

Table 6.1. K-Ar data (from Aronson and Hower 1976)

depth (m)	K_2O (%)	^{40}Ar rad./g	Apparent age (Ma)
shale:			
1250 m	2.03	4.70	150.5±4
2450 m	2.40	4.57	124.5±4
3100 m	2.06	4.15	132.0±4
3400 m	2.24	4.11	120.0±4
3550 m	2.27	3.82	110.5±4
3700 m	3.13	4.34	91.5±3
4000 m	3.78	4.93	86.0±2
4300 m	3.48	3.97	75.5±2
4600 m	3.10	3.73	79.5±2
4900 m	3.31	3.51	70.5±2
5200 m	3.46	4.01	77.5±3
5500 m	3.07	3.28	71.0±3
<0.1 μm clay fractions:			
1250 m	2.26	1.830	53.9±2
1850 m	1.81	1.500	55.2±2
2150 m	2.11	1.745	55.2±4
3100 m	2.55	1.755	45.9±3
3400 m	3.17	2.185	46.1±2
3550 m	3.29	2.285	46.4±2
3700 m	4.62	2.360	34.2±2
4000 m	5.27	2.735	34.8±2
4300 m	5.27	2.560	32.6±2
4900 m	5.28	2.390	30.4±2
5200 m	4.88	2.385	32.8±2
5500 m	4.84	2.355	32.6±2

The isotope study of Aronson and Hower (1976) makes it clear that the K-Ar ages of detrital sediments have to be considered with great care and that a tentative interpretation is only possible where the mineralogic and geochemical basis of the rock system has been demonstrated. It is improbable that diagenetic clay minerals form in isotopic equilibrium with detrital clay minerals (Hamilton et al. 1989).

If it is not possible to sample pure fractions of authigenic clay minerals, then producing mixed ages is unavoidable. It is just as difficult in the case of newly formed clay minerals. Recrystallization is characterized by the break up and new formation of chemical bonds. As we can scarcely decide to what extent all pre-existing bonds have been broken or how many crystallographic cells of the precursor mineral have remained intact through the recrystallization process, it is impossible to rule out K-Ar mixed ages.

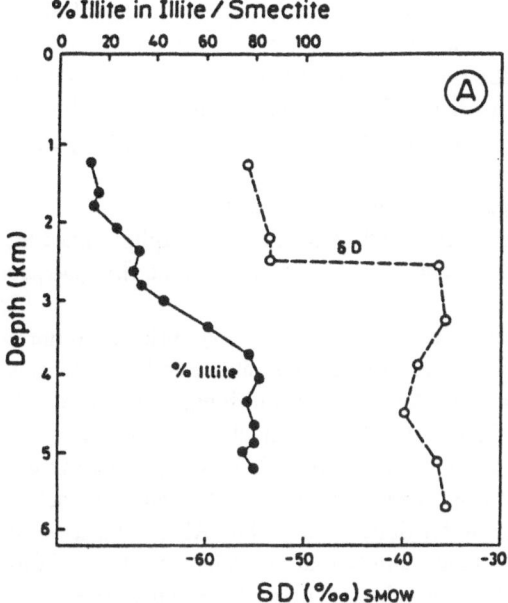

Fig. 6.8. δD and percent illite as a function of depth. (Yeh 1980)

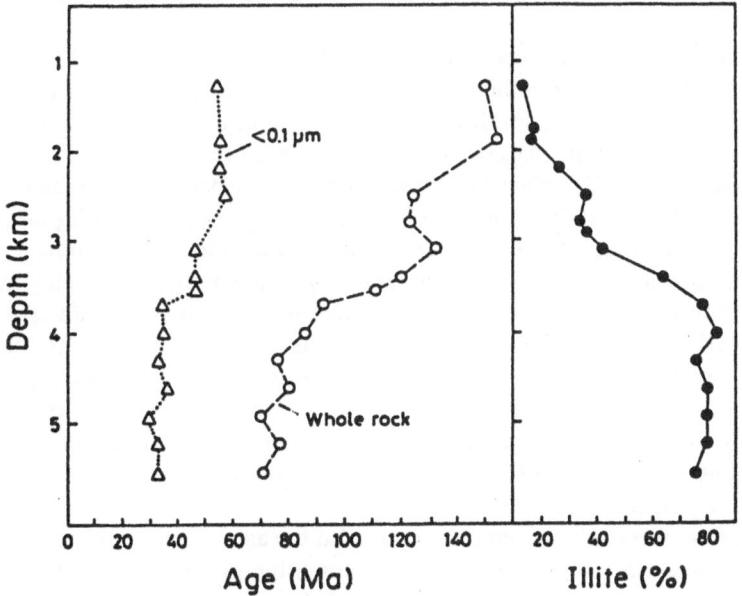

Fig. 6.9. Depth dependence of K-Ar ages. (Aronson and Hower 1976)

This is made clear in the study of Clauer et al. (1990), who investigated recrystallized smectite from lower Cretaceous and Paleocene shales from the North Atlantic, not only isotopically (Rb-Sr, O, D) but also using electron microscopy, X-ray diffraction and major and trace element analyses. Three different populations of smectite could be differentiated by morphology :

A) a flake-like type with diffuse and irregular rims,
B) a lath-like type, well crystallized with idiomorphic rims and,
C) an intermediate type with similarites to both the first and second types.

The lath-like type was found to be particularly enriched in the finest fraction (<0.2 µm), while the flakes were found most abundantly in the coarsest fraction. The authors suggest that the laths are of authigenic origin and the flakes are more likely to be detrital smectite. The chemical investigations allow no differences to be established between the three kinds. Likewise, rare earth elements, normalized to the standard (average North American shales) are not fractionated and yield the characteristic patterns for detrital clay. The REE data, as well as the major and trace element information, imply that the laths were formed directly from the pre-existing flakes, without any loss or gain of elements. The process appears to have taken place according to the rule of dissolution/precipitation.

Various grain-size fractions separated out from the clays were brought into suspension over 15 minutes in 1N HCl and allowed to react with the hydrochloric acid (leaching). The Sr isotopic compositions of residues and leachates arising from this experiment were analyzed.

According to mass balance laws, untreated clay fractions as well as leachates and residues deriving from this sort of experiment should all lie on a mixing line on a Rb/Sr isochron diagram. If the clay minerals formed in isotopic equilibrium with the ambient fluid phases, from which other accessory minerals possibly crystallized, then this straight line must represent an isochron. Its gradient should reflect the age of crystallization, i.e. of isotopic equilibrium and of diagenesis. The inital Sr isotope ratio of this isochron ought to reflect the isotopic composition of the environment in which the clay minerals crystallized. Should the clay minerals also contain old, inherited, detrital parts, then these gradients would represent a geologically meaningless age and the straight line would reflect mixing between detritus and diagenetic new formation. The results of the leaching experiments define straight lines in isochron diagrams whose gradients are directly related to the grain size (Fig. 6.12). The finer the grain size, the lower are the calculated, apparent ages. These correlation lines represent mixing lines between the detrital flake and authigenic lath types.

Therefore, these mixing lines do not record the age of formation. The decrease in the steepness of the mixing line with decreasing grain size records either the preferential transfer and transport away of ^{87}Sr into the porewater or may be a result of an increase in the Rb/Sr ratio in the fine, mostly authigenic, clay fraction. The Sr isotopic ratios do not reflect the isotopic composition of seawater, and so demonstrate, as do the oxygen isotope ratios, that this new mineral formation has not taken place in contact with seawater. The various grain sizes represent isotopic mixtures and direct age dating is not possible. Clauer et al. (1990) showed however one possibility for indirect dating in their work. They compared the apparent ages with the relative amount of lath-type crystals (Fig. 6.12).

If these mixing lines are thus extrapolated to relative proportion of laths'=100%, then the approximate age of recrystallization can be deduced. These extrapolated ages all lie between 100 and 60 Ma and are practically identical to the likely stratigraphic ages.

These studies make it plain that reconstructing the age of recrystallization of detrital sediments using the Rb-Sr or the K-Ar methods is extremely tricky but that these methods can, on the other hand, provide us with valuable information regarding sediment petrogenesis. In Chap. 7, we will look once again in detail at the problems of dating detrital sediments, using the Sm-Nd isotopic system.

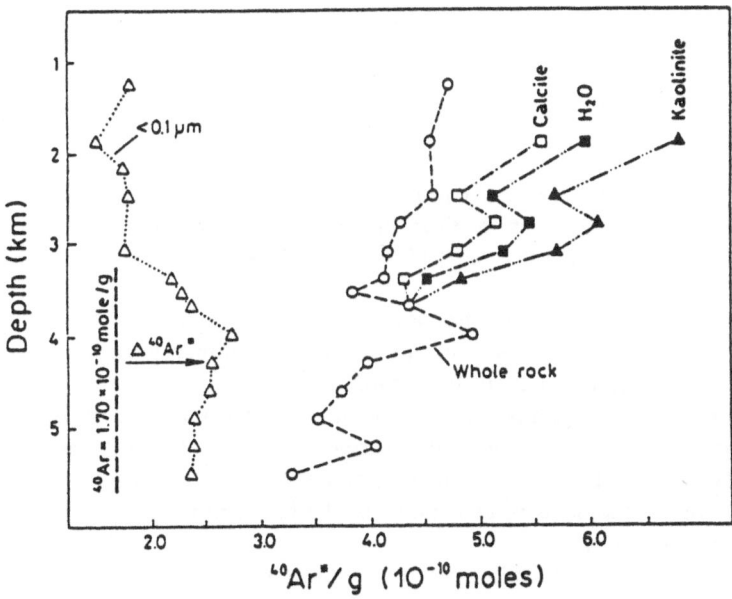

Fig. 6.10. Depth dependence of rad. argon ($^{40}Ar^*$) in <0.1µm clay-size fractions and whole shales. (Aronson and Hower 1976)

6.2 Glauconite Formation

As glauconite samples are easier to concentrate than other clay minerals and because the quality and purity of the glauconite fraction can be ascertained using a simple binocular microscope, it is easier to investigate them isotopically and geo-

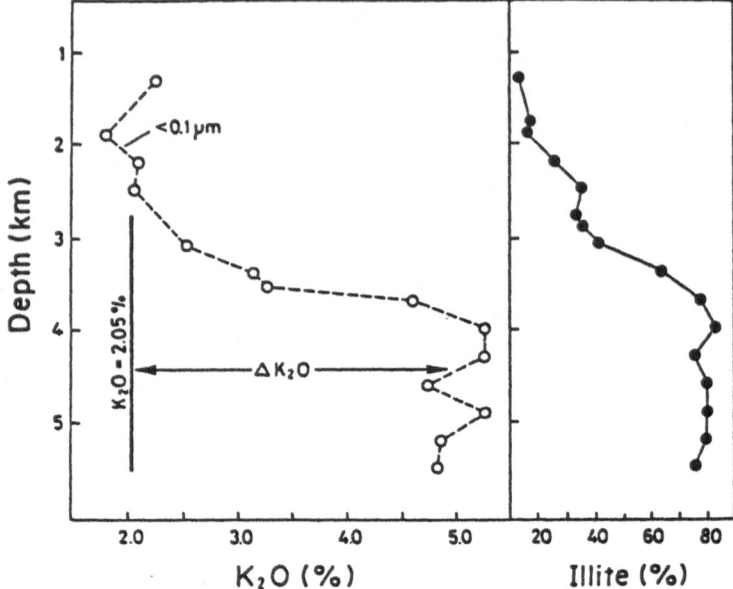

Fig. 6.11. Potassium concentrations in the <0.1 μm clay size-fractions as a function of depth. ΔK₂O is the additional K_2O in the samples below 3000 meters relative to concentrations in the shallower samples. (Aronson and Hower 1976)

chemically than the <2 μm clay fraction, which beside authigenic clay minerals is likely to contain a small proportion of detrital clay minerals or other minerals, whose identification is not at all straight forward.

Glauconites are iron and potassium rich clay minerals that may be of authigenic or of detrital origin. They form in the marine environment during periods of low sedimentation rate. Detailed information about what we know today concerning the formation mechanisms of glauconite can be found in the book "Green marine clays" by Odin (1988).

While authigenic glauconites may incorporate seawater isotopic signatures as they formed in chemical equilibrium with seawater, detrital glauconites show strongly divergent isotope characteristics.

Authigenic glauconites allow isotope age determinations which define not only the age of formation but also the isotopic composition of coeval seawater. Rb-Sr and K-Ar isotopic ages determined on glauconites that originally formed from detritus remain extremely difficult to interpret. The proportion of inherited, radiogenic [40]Ar and [87]Sr is just as difficult to establish as in the previously discussed cases.

The genesis of detrital glauconites is still controversial today. For example, Fischer (1987) showed that sedimentary biotite which has been reworked may recrystallize to glauconite through chlorite at or close to the sediment/water interface. A much debated form of originally detrital glauconite is that of 'worm' faecal pellets (Odin and Matter 1981). The genesis of these glauconites is discussed in a little more detail here:

Glauconites of this type have been investigated from sediments in the Gulf of Guinea. This region is particularly suitable because it is possible to observe various stages of glauconitization and neither strong subsidence nor high sedimentation rates have disturbed the progressive evolution of these glauconites in the sediment / seawater transition zone.

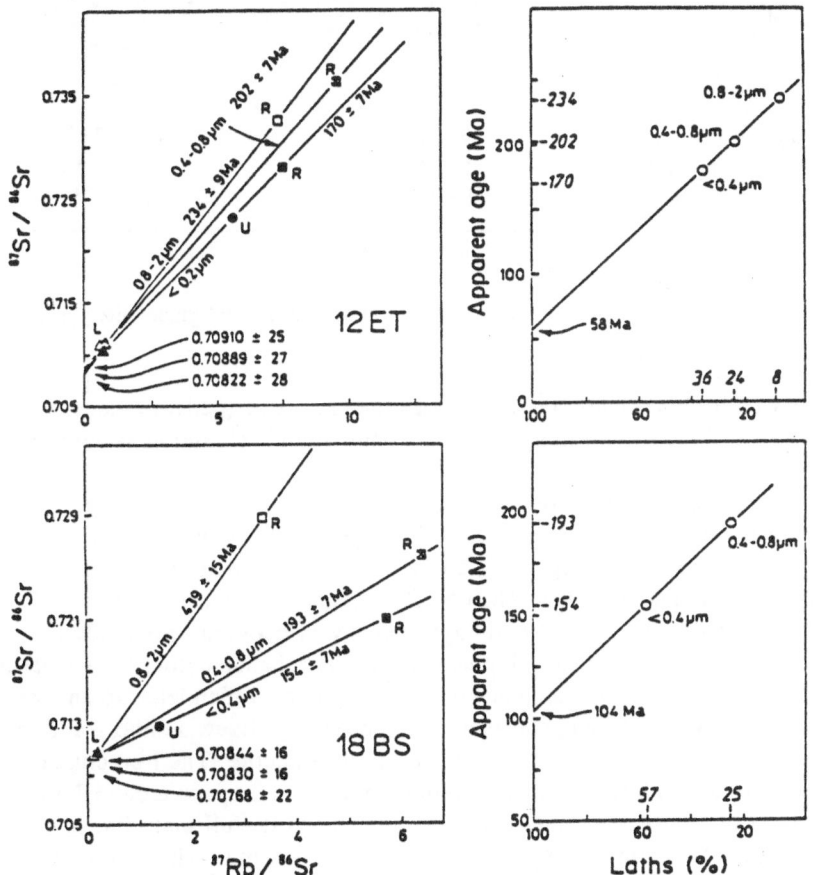

Fig. 6.12. Rb-Sr isochron diagrams and apparent Sr ages of recrystallized smectites from lower Cretaceous and Paleocene shales. The slopes of the reference lines are a function of the size-fractions. L leachates, R residues, U untreated sample. Comparison of the apparent ages with the relative amount of lath-type crystals and extrapolation to relative proportion of laths = 100% allows us to calculate the approximate age of recrystallization. (Clauer et al. 1990)

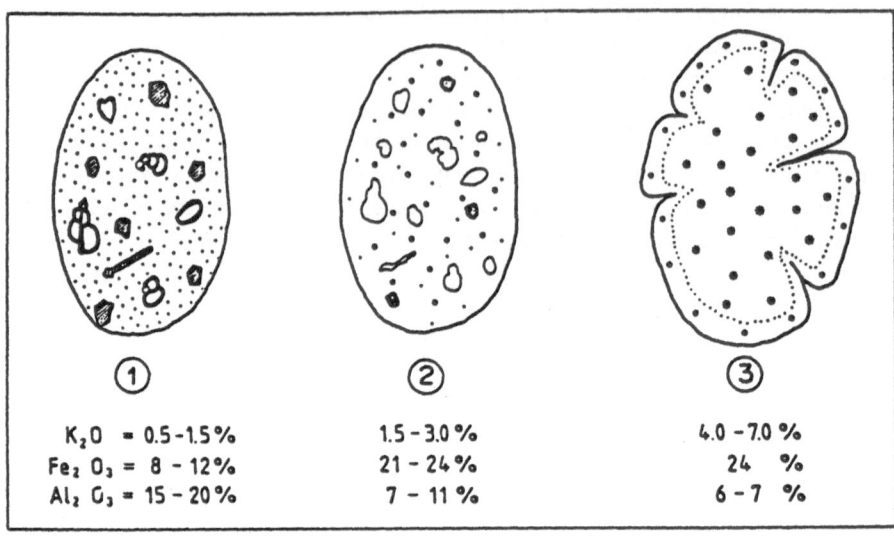

Fig. 6.13. The morphologic and mineralogic evolution of pelletal glauconies. (after Odin 1988) ·

Before we turn our attention to the isotope data for these (pelletal) glauconites we ought first to look at their morphologic and mineralogic evolution as described by Odin (1988) from the Gulf of Guinea:

In stage 1 of glauconite formation, these pellets consist of sedimentary residue and biogenic particles. These pellets, which represent an initial stage of glauconite formation, are shown in Fig. 6.13, and are made up of detrital kaolinite, coarse quartz and carbonate residues. Their K_2O contents lie between 0.5 and 1.5%, the Fe_2O_3 contents 8-12% and their Al_2O_3 contents vary between 15 and 20%.

During stage 2 of their development the pellets change colour somewhat to a light green. The biogenic components go into solution which leads to an increase in porosity. The quartz and carbonate components go likewise into solution and disappear. The kaolinite begins to transform into smectite. The iron content rises to between 14% and 21% while the aluminium content drops to around 7-11%.

Stage 3 of this process sees the glauconite reach its full maturity. It loses its original pelletal form and begins to increase in volume. The grains leave no recognizable detrital relicts or shell fragments. This allows us to assume that the detrital components have all been destroyed without trace and complete recrystallization has been completed. Now these are no longer detrital mud-faecal pellets, instead they represent mature glauconite grains. The potassium content has risen again and now varies between 4 and 7%, while the Fe and Al contents do not change very much during this final stage.

How can the development of these glauconite grains be represented using the various isotope systems? Oxygen isotope data from four glauconite samples, which represent stages 2 and 3, respectively, are displayed in Table 6.2. The K concentrations of the samples G313A and G490h reflect stage 2 of the glauconite maturation process. The low $\delta^{18}O$ values of 18.23 (SMOW) show clearly the presence of detritus in the pellets and allows us to suppose that little or no isotopic exchange has taken place with seawater at this early stage of development. This is worth noting because according to X-ray diffraction, the kaolinite structure has already been destroyed without trace and smectite has fully replaced it.

Table 6.2. $\delta^{18}O$ values, K-Ar data and apparent ages of glauconites. (after Odin 1988)

sample	age of sedimen-tation (years)	K_2O %	$\delta^{18}O$	Ar rad. (nl/g)	age (Ma)
G313A	16000	3.0	18.23	15.0	150±10
G490h	20000	3.4	20.80	8.9	·80±7
G490 e	20000	4.2	21.48		
G490a	20000	4.25	21.10		
G490e+a	20000	4.2	7.1		50±3

Obviously, these early glauconites represent closed systems where oxygen can neither be taken in or released. We can see that the $\delta^{18}O$ values also rise with increasing degree of maturity and thus K concentration. Even at K concentrations of 4.25%, however, we can see that with $\delta^{18}O$ values of 21.1-21.5, isotopic equilibrium has not yet been reached. For glauconites in equilibrium with seawater we would expect values between 23 and 25‰ SMOW.

K-Ar ages from these glauconites can provide us with much information. The clay mud, in which the pellets and the glauconites are found, yields ages of between 520 and 470 Ma (Fig. 6.14). The clay minerals in the mud, which most probably derive from surface weathering on the African continent and were transported into the Congo river, are as we found out in Chap. 3, rich in radiogenic Sr and Ar and are responsible for such a high apparent age.

The pellets and the glauconites show decreasing apparent ages with increasing amount of maturity (increasing K contents). But even the sample most enriched in potassium (7.5%), has not yet reached its stratigraphic age of about 1 million years and instead shows an age of some 11 Ma (Fig. 6.14). The higher the relative content of potassium, the lower the apparent age. This relationship can be attributed either to the loss of radiogenic ^{40}Ar and/or the addition of K during glauconite formation.

In the case of addition of potassium, the newly forming glauconite grains must have contained a significant amount of inherited, radiogenic ^{40}Ar. Odin and Matter (1981), Odin and Dodson (1982) and Odin and Fullagar (1988) postulate on the basis of their data a two-stage "open system" model to explain the process of glauconitization:

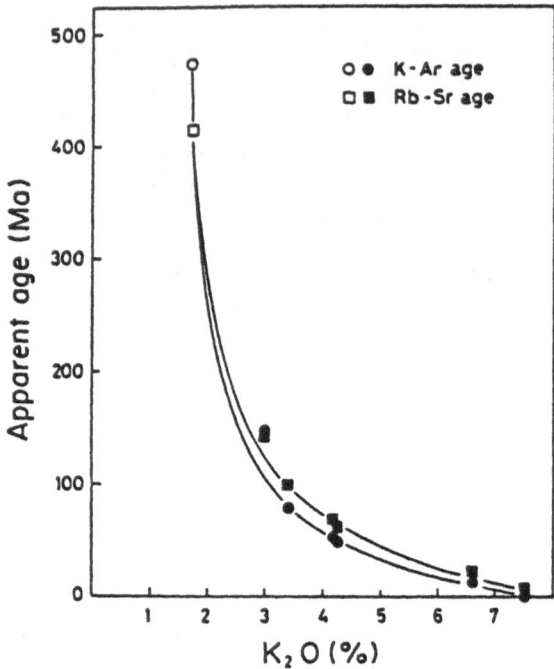

Fig. 6.14. Variation of K-Ar and Rb-Sr apparent ages in glauconies (filled symbols) and clays (open symbols) after Clauer et al. (1992)

The first phase of glauconitization is characterized by the growth and new formation of smectite in the pores of the faecal pellets, whereby cations from the sea and pore waters can be drawn upon. The second phase corresponds to recrystallization, the glauconite forms, grows further, taking up potassium from its environment.

A combined Sr-Nd isotope study, carried out on the same glauconite samples, allows us to record a far more differentiated picture of glauconite genesis and shows that glauconites were closed systems during their earlier stages of development and only became more open during later glauconite growth when they began to exchange with seawater (Clauer et al. 1992; Stille and Clauer 1994). The Rb-Sr model ages of recent glauconites are compared with their K-Ar apparent ages in Fig. 6.14. The Rb-Sr model ages were calculated under the assumption that glauconites which have undergone exchange with seawater of today's Sr isotopic composition must yield an age of zero. The Rb-Sr isotope system behaves very similarly to the K-Ar system. The Sr model ages drop with rising K content but without reaching the zero age.

HCl leaching experiments carried out on the glauconites make it clear that this decrease in the Sr model ages and the K-Ar apparent ages are not the result of the loss of ^{87}Sr and ^{40}Ar, respectively, in an open system, but are related to a rise in the Rb and K concentrations in a closed system (Fig. 6.15).

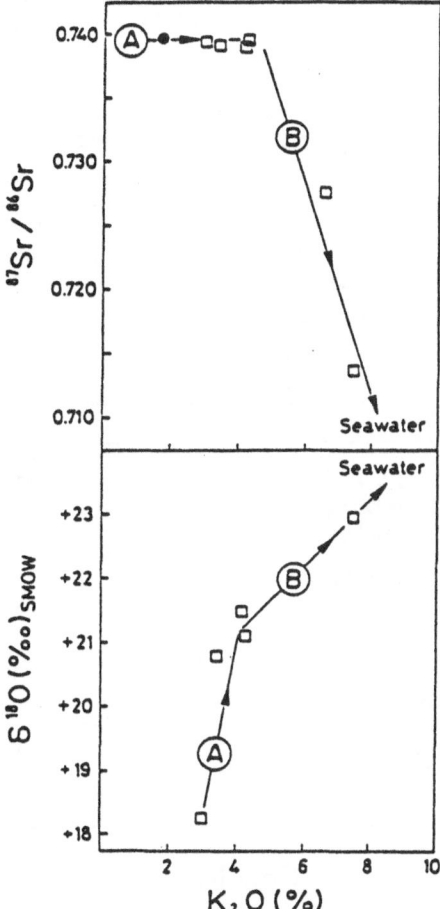

Fig. 6.15. Relationship between $^{87}Sr/^{86}Sr$, $\delta^{18}O$ and K_2O during the evolution of glauconies. Glauconies (squares) and associated clay (filled circle). (after Clauer et al. 1992)

Thus, 1N HCl leachates of the detrital clay from which the glauconites formed, show Sr isotopic compositions which are identical with those of the glauconite residues (K_2O <4.5%) with model ages between 410 and 60 Ma and K-Ar apparent ages between 470 and 50 Ma (arrow A in Fig. 6.15).

This observation implies that there is a close genetic relationship between the clay mud and the glauconites and that this relationship is retained into stage 2 of glauconite formation as no radiogenic ^{87}Sr has been lost and also not even minor exchange with seawater strontium has taken place by then. Thus, apparently contradicting the opinions of Odin and Fullagar (1988), the glauconites appear to have remained a closed system.

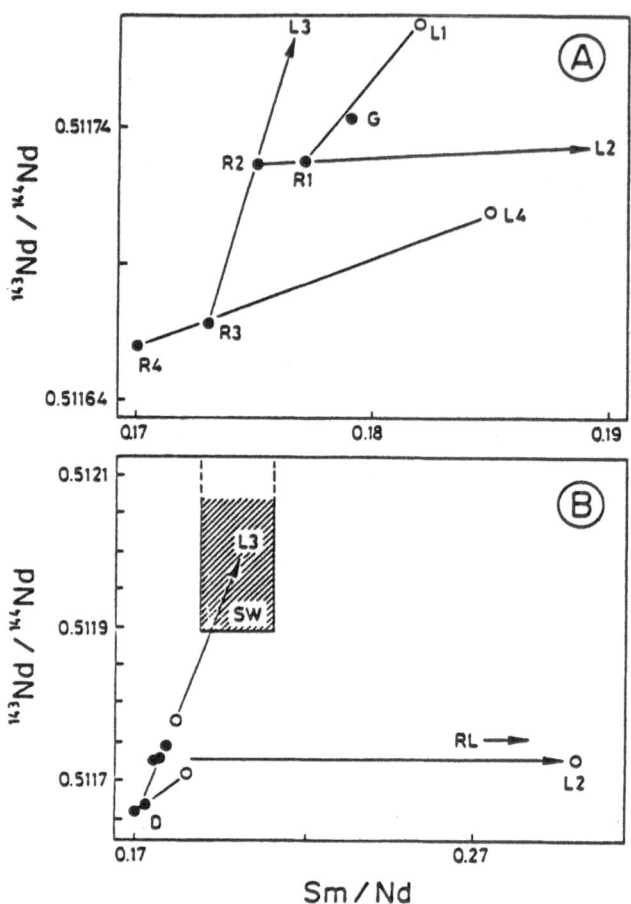

Fig. 6.16. Sm-Nd isotopic analyses of glauconite residues (R1-R4) and leachates (L1-L4) show that even glauconies with a K_2O content greater than 4% are isotopically not homogeneous. A: enlargement of Fig.19B. Three isotopically different components are recognizable: D detritus, SW seawater, RL suspension. (Stille and Clauer, 1994)

Only in growth stage III, (arrow B in Fig. 6.15) does the Sr isotopic composition begin to fall and agree with that of seawater. To what extent does the glauconite system open itself to exchange with seawater?

Results of several complementary leaching experiments are of particular importance here. These experiments were treated first with water (L1), then the resulting residue with 0.1 N acetic acid (L2) and thereafter twice with 1N HCl (L3, L4). This experiment was designed to leach ever deeper into the glauconite structure and to break it up. The chemical analyses, as well as the isotopic ratios of these acid leachates, allow us to identify chemically and isotopically various components, which have possibly been derived from detritus and have since remained intact. Sm-Nd isotopic studies of these glauconite residues (R1-R4) and leachates (L1- L4) show that even glauconite with a K_2O content of greater than 4% is not isotopically homogeneous (Fig. 6.16). At least three isotopically distinguishable components can be observed (D, SW, RL). While the Nd isotopic compositions of the leachates are strongly variable, Sr isotopic ratios show no variation and are comparable with the Sr isotope composition of seawater (Fig. 6.17).

Component D' is characterized by the glauconite residue and represents the detrital proportion in the glauconite. The HCl-1 acid leachate L3 is strongly enriched in phosphate (P_2O_5 >1.5%) and the REE. Its Nd isotope composition is identical to that of seawater. It is thus suggested that component SW corresponds to a phosphate-rich phase that formed in Sr and Nd isotopic equilibrium with seawater. This phase may consist of colloidal apatite or apatite inclusions within the glauconite (Fischer and Steiger 1988; Odin 1988).

The HAc acid leachate L2 and the HCl-2 leachate L4 show no phosphate and are strongly enriched in calcium and organic carbon. Their Sr isotope compositions are identical to seawater but their Nd isotopic compositions lie somewhere between continental detritus and seawater. Stille and Clauer (1994) suggest that leachates L2 and L4 reflect calcium and organic carbon rich phases in

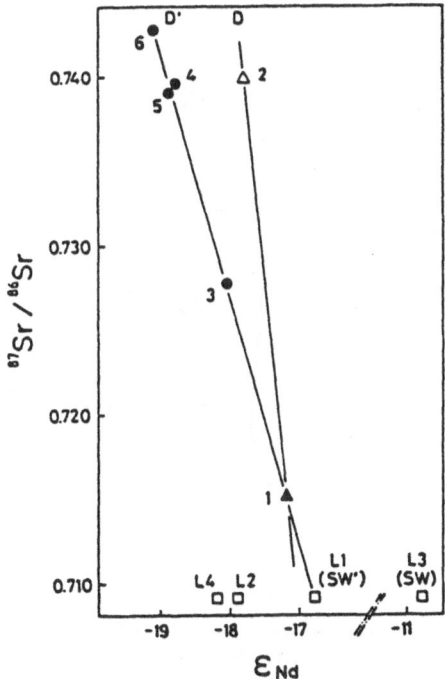

Fig. 6.17. $^{87}Sr/^{86}Sr$ ratio versus εNd of: (1 untreated clay fraction, 2 clay residue after leaching in 1N HCl: L leachates: L1=H_2O, L2=0.1NHAc, L3=1N HCl, L4=1N HCl. (Stille and Clauer 1994)

suspension in the Congo River, respectively, which ended up deposited in the river delta (component RL in Fig. 6.18). Contact with Sr-rich and REE-poor seawater would overprint the Sr isotopic composition in particular but much less the Nd isotopic signature of the carbonate and the organic components.

Components D, RL and SW are shown schematically in Fig. 6.18 in a Sr-Nd isotope diagram. The glauconites move inside the triangle D-RL-SW depending on their respective degrees of maturity (see potassium contents) along a straight line (SW'-D'; see also Fig. 6.17). The two points of transection of the triangle edges along this line determine the mixing end members DRL and SW'. SW' is defined by the relative proportions of suspended load (RL) and phosphate-rich phase (SW) in the glauconite, while DRL is derived from the mixing of a detrital silicate phase (D) with suspended load (RL). Component DRL represents the mud in which the worm lived and which he consumes. These components all originate therefore from the faecal pellet, which was the initial stage of glauconitization.

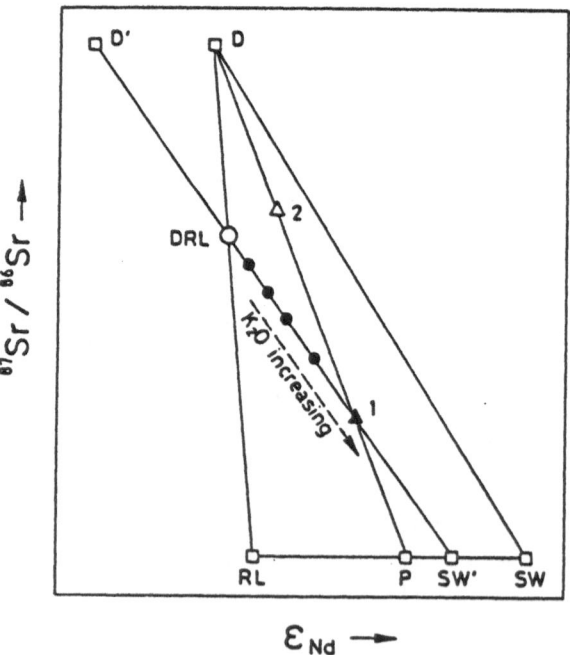

Fig. 6.18. Schematic Sr-Nd isotope diagram deduced from Fig. 6.17 (D continental detritus, silicates, RL carbonates and organic river load, SW authigenic phosphates with seawater isotopic signatures, DRL mud, 1 clay of the mud, 2 clay after HCl treatment). For explanation of P, SW' and D' see text. The filled circles represent glauconies; the arrows are indicating the direction of K_2O increase in the untreated glauconies. The solid lines D'-SW' and D-P are drawn as straight lines for simplicity (Stille and Clauer 1994).

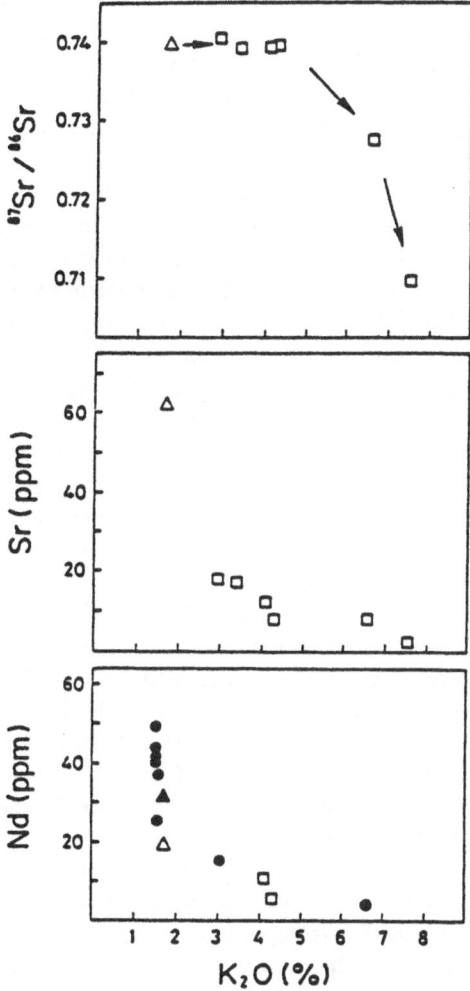

Fig. 6.19. $^{87}Sr/^{86}Sr$ ratios and Sr concentration values of glaucony residues (squares), muddy clays (triangle) and untreated glauconies (filled circles) relative to their K_2O contents. (Stille and Clauer 1994)

Armed with this information, we are now in a position to develop a model of glauconitization. In the early stages of glauconite genesis (K_2O <4%), Sr and Nd are released from the glauconite. Its isotopic compositions are identical to those of the Sr and Nd remaining in the crystal lattice and with the Sr and Nd isotopic compositions of the clay mud in which the worm lived. This is made clearer in Fig. 6.19 in which the Sr and Nd concentrations and Sr isotopic ratios are shown relative to K_2O contents. This first stage of glauconite formation is therefore dominated by the chemical and isotopic signature of the glauconite precursor (component DRL in Fig. 6.18). Dissolution and crystallization processes must have run their course during this time without the opening of the glauconite system to the outside and with seawater. It can therefore be excluded from consideration that the rise in potassium content reflects any input of potassium. It is possible that this apparent increase is due to volume changes only, which are characteristic for this stage.

The $^{87}Sr/^{86}Sr$ ratios begin to fall (Fig. 6.19) while the $^{143}Nd/^{144}Nd$ ratios begin to rise around a K content of about 4.5%. Thus, more mature glauconites start to move along the D'-SW' mixing line (Fig. 6.17 and 6.18). As the Sr and Nd concentrations continue to fall (Fig. 6.19), the decrease in the $^{87}Sr/^{86}Sr$ ratios and the increase in the $^{143}Nd/^{144}Nd$ ratios can only be explained by the preferential release of radiogenic Sr and non-radiogenic Nd of component D from the glauconite. Dissolution of the now chemically unstable, detrital clay fraction is the most likely mechanism responsible for the loss of Sr and Nd. The increasing loss of these elements from the detrital fraction has the consequence that the influence of Sr and Nd rich phosphatic components (SW) is greatly increased. As the glauconites start to grow at this stage of their development it can be assumed that chemical equilibrium and isotopic exchange with the immediate environment also begins at this point. This direct exchange with seawater could possibly have led to the decrease in the $^{87}Sr/^{86}Sr$ ratios. As REE concentrations in seawater are comparatively low, it can be assumed that exchange is not responsible for the rise in Nd isotope ratios, whilst remembering that both Sr and Nd contents decrease with increasing glauconite development and K_2O content. Stille and Clauer (1994) suggest that the reestablishment of a seawater Sr and Nd isotope signature in the glauconite is largely controlled by the amount of Sr and Nd in detrital components that have been released and through the amount and Sr and Nd contents of phosphate-rich phases in the glauconite.

These studies make it clear that the isotopic composition of detrital clay minerals can be altered during recrystallization to suit the environment of deposition or formation, but that isotopic equilibrium is unlikely to be reached. The possibilities for dating such minerals are therefore restricted. However, authigenic glauconites, which formed in chemical equilibrium with seawater and have incorporated seawater isotopic composition, can be treated differently.

6.3 Dating of Authigenic Clay Minerals

Rb-Sr isotope work on authigenic glauconites from the Island of Crete containing up to 8% potassium were used for fixing the age of the Cenomanian / Turonian boundary by Odin and Hunziker (1982). The samples define a straight line whose gradient defines an age of 93.5 ± 1.6 Ma (Fig. 6.20). The age agrees with K-Ar apparent ages (93.0 ± 1.4 Ma) and also with the assumed stratigraphic age.

As the coexisting smectites fall likewise on a glauconite isochron, it can be assumed that at the time of the closing of the Sr system, isotopic equilibrium had been reached between smectite and glauconite. The inital Sr isotopic composition of 0.708 is slightly higher than the likely $^{87}Sr/^{86}Sr$ of seawater 93 Ma ago. This might imply that the Sr system closed after the glauconites became isolated from the marine milieu. Other examples of absolute dating of glauconites are found in the book of Odin (1982).

Rb-Sr age determination can also be applied to authigenic smectite and palygorskite, if concentrates can be isolated cleanly enough. A study of Clauer (1976) illustrates this possibility, in which an attempt was made to date Paleocene, detrital sediments of the Mormoiron basin in the southeastern part of France using the Rb-Sr dating technique. Smectite and palygorskite fractions were concentrated and gave Rb-Sr isochron ages of 59.1 ± 1.4 Ma (Fig. 6.21), which agrees well with the expected stratigraphic age for this part of the Paleocene.

Clay mineral fractions, which are enriched in authigenic illite/smectite are commonly used to date diagenetic processes using the Rb-Sr method. However, it is apparent that even using the most modern method for concentrating clay mineral fractions it is not always possible to separate the authigenic from the detrital clay minerals (e.g. Liewig et al. 1987). The influence of a detrital component on the Sr isotopic composition of the authigenic phase is displayed in Fig. 6.22. In this commonly occurring case-model, the Sr isotope composition of the detrital component (DC) is far higher than that of the authigenic phases that are in equilibrium with each other. It is further assumed that authigenic and detrital phases are mixed with each other in approximately the same proportion.

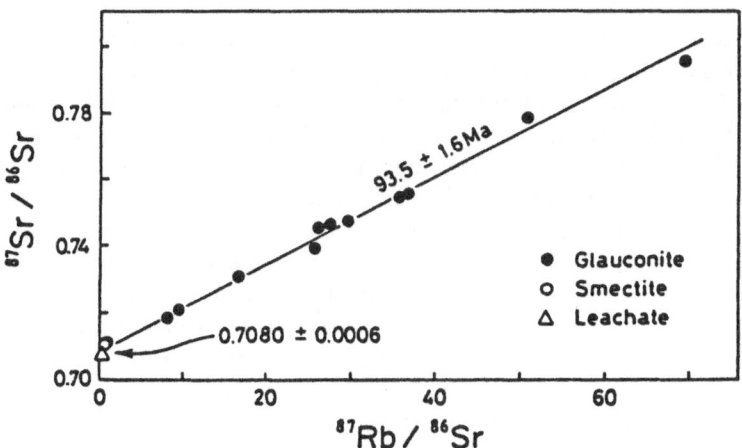

Fig. 6.20. Rb-Sr isochron diagram for authigenic glauconites and smectites from the Island of Crete. The Sr age agrees well with K-Ar isotope determinations. (Odin and Hunziker, 1982)

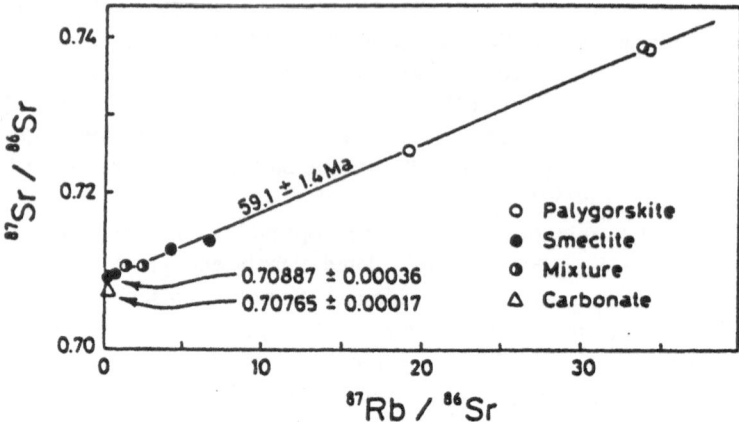

Fig. 6.21. Rb-Sr isochron diagram for clays from a Tertiary lacustrine sedimentary sequence in southeastern France. The Sr age agrees with the expected stratigraphic age. (Clauer 1976)

The presence of the detrital component DC has the consequence that every authigenic sample point lying on a mixing line (DC-1, DC-2, DC-3, DC-4) will be moved in the direction DC and so move ever further away from the isochron. In the case that all samples are mixed with detritus to the same extent, these sample points will also lie on a straight line but above the isochron. Only by very minor mixing with detritus will such a line show a gradient and age that agrees with the age of diagenesis. With increasing addition of detritus, the gradient will rise sharply, thus increasing the subsequent age obtained, an age which will be geologically irrelevant.

The Rb-Sr isotope study of Gorokhov et al. (1994) is of interest here. In this study, various illite grain-size fractions (0.6-2.0 µm, 0.4-0.6 µm, 0.3-0.4 µm, 0.2-0.3 µm, 0.1-0.2 µm) were analyzed from samples close to the Precambrian / Cambrian boundary in the northwestern part of the East European Platform. These samples contained varying proportions of detrital clay minerals. The various fractions were leached and these leachates were analyzed isotopically. From this work it could be seen that with decreasing grain-size, there was also a lowering of the apparent age of the sample (Fig. 6.23). The apparent age of the coarsest clay mineral fraction was 722 ± 13 Ma while the finest fraction provided an age of 533 ± 8 Ma. The Rb-Sr apparent ages fall therefore by 200 million years with decreasing grain-size. Only by carrying out investigations into the clay mineralogy of the sample material can we attempt to interpret these isotope ages. The fine fractions are enriched in the 1M-1Md illite polymorphs (i.e. likely to be authigenic). Their presence permits the assumption that the clay mineral rich rocks have never reached characteristic temperatures for low-grade metamorphism and have thus remained within the normal temperature range for diagenesis. The 2M illites (detrital) in the coarse fractions are characteristic of low-grade metamorphism and permit us to suppose that these represent detrital grains.

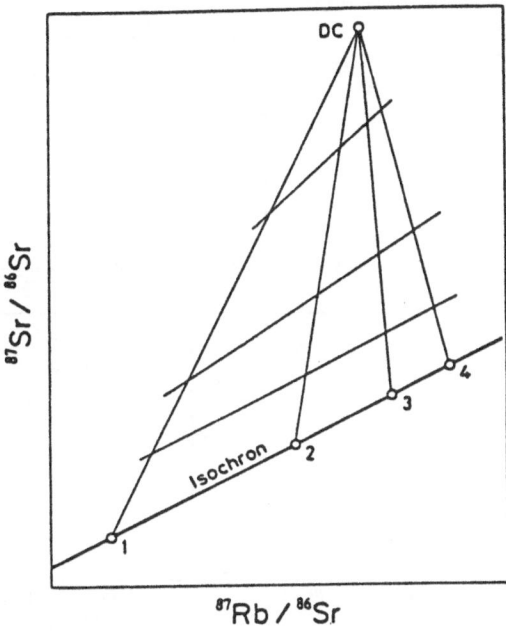

Fig. 6.22. Rb-Sr isochron diagram illustrating schematically the influence of a detrital component (DC) on the Sr isotopic composition of the authigenic clay mineral phases 1 to 4, which are in isotopic equilibrium (modified after Awwiller, 1994). It is assumed that authigenic and detrital phases are mixed with each other in approximately the same proportion (condition to get another straight line, or errorchron without age significance). Admixture of high amounts of the detrital component DC strongly increases the slope of the corresponding errorchron (for further explanation see text).

The clay mineral fractions merely represent mixtures of illites of authigenic origin (1M-1Md) and those of detrital origin (2M). From these results the authors surmise that the age of 722 ± 13 Ma, found from the 0.6-2.0 µm fraction must represent the time of formation of the first, detrital illite generation, whereas the age of 533 ± 8 Ma, found from the finest fraction, reflects the time of diagenesis and formation of the second, authigenic illite generation. This age is considered to represent a minimum sedimentation age for the lower "pretrilobite" Cambrian and has been shown by several recent studies to be a reasonable age on the basis of single zircon U/Pb dating of tuffs (e.g. Grotzinger et al 1995). The ages calculated from the intermediate 0.6-0.2 µm grain-size fraction is not geologically relevant. On the basis of the grain-size dependency of these apparent ages, one would expect that the clay mineral fraction <0.1 µm should provide even lower ages. For this third illite generation, ages of between 481 and 466 Ma are derived. The authors surmise that this illite age represents a phase of retrograde katagenesis (late-stage diagenesis) that can indeed be recognized across the whole region under study and which overprinted these sediments between 480 and 430 Ma ago.

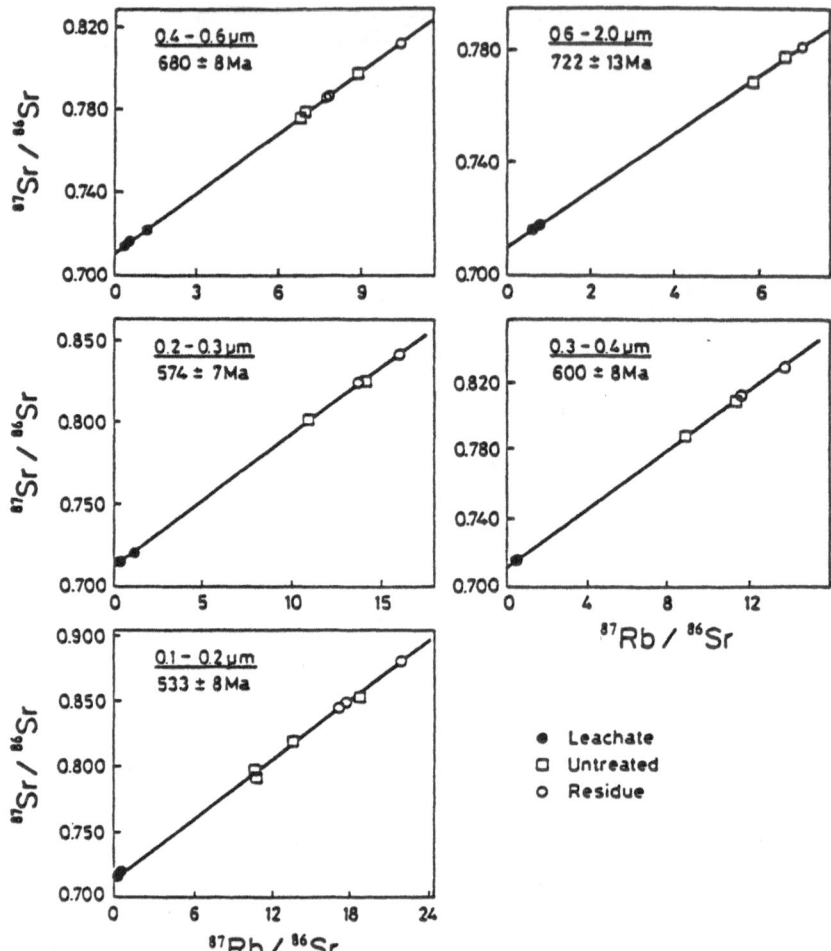

Fig. 6.23. Rb-Sr isochron diagrams for various illite grain-size fractions from samples close to the Precambrian/ Cambrian boundary. The clay mineral fractions represent mixtures of authigenic and detrital illites. The clay fractions 0.1-0.2 μm mainly contain authigenic clay minerals and define an age of 533 ± 0.2 Ma. This age is considered to represent a minimum sedimentation age close to the Precambrian-Cambrian boundary, (after Gorokhov et al. 1994)

This investigation shows once more that the interpretation of Rb-Sr isotope data on clay mineral fractions is problematic because of the difficulty of separating authigenic clay mineral fractions from detrital ones. Only accompanying studies of the detailed clay mineralogy and general geology can help us towards a satisfactory interpretation of isotopic ages.

6.4 References

Awwiller DN (1994) Geochronology and mass transfer in Gulf Coast mudrocks (south-central Texas, U.S.A.): Rb-Sr, Sm-Nd and REE systematics. Chem Geol 116: 61-84

Aronson JL, Hower J (1976) Mechanism of burial metamorphism of argillaceous sediment: 2. Radiogenic argon evidence. Geol Soc Amer Bull, 87: 738-744

Biscaye PE (1965) Mineralogy and sedimentation of Recent deep-sea clay in the Atlantic Ocean and adjacent seas and oceans. Geol Soc Amer Bull, 76: 803-832

Clauer N, O'Neil JR, Bonnot-Courtois Ch,. Holtzapffel Th (1990) Morphological, chemical, and isotopic evidence for an early diagenetic evolution of detrital smectite in Marine sediments. Clays and Clay Minerals, 38: 33-46

Clauer N, Chaudhuri S, Kralik M, Bonnot-Courtois Ch (1993) Effects of experimental leaching on Rb-Sr and K-Ar isotopic systems and REE contents of diagenetic illite. Chem Geol, 103: 1 -16

Clauer N, Stille P, Keppens E, O'Neil JR (1992) Le mécanisme de la glauconitisation: apports de la géochimie isotopique du strontium, du néodyme et de l'oxygene de glauconies récentes. C R Acad Sci Paris, 315: 321-327

Dasch EJ, Dymond JR, Heath GR (1971) Isotopic analysis of metalliferous sediment from East Pacific Rise. Earth Planet Sci Lett, 13: 175-180

Dasch EJ (1969) Strontium isotopes in weathering profiles, deep-sea sediments, and sedimentary rocks. Geochim Cosmochim Acta, 33: 1521-1552

Fischer H (1987) Excess K-Ar ages of glauconite from the Upper Marine Molasse and evidence for glauconitization of mica. Geol Rund, 76/3: 885- 902

Fischer H, Steiger, RH (1988) The influence of sediment lithification on K-Ar ages and chemical zoning of glauconites. Schweiz Min Petr Mitt, 68/2: 201 - 212

Gorokhov IM, Clauer N, Turchenko TL, Melnikov NN, Kutyavin EP, Pirrus E, Baskakov AV (1994) Rb-Sr systematics of Vendian-Cambrian claystones from the east European Platform: implications for a multi-stage illite evolution. Chem Geol, 112: 71-89

Grotzinger JP, Bowring SA, Saylor BZ, Kaufman AJ (1995) Biostratigraphic and geochronologic constraints on early animal evolution. Science, 270: 598-604

Hamilton PJ, Kelley S, Fallick, AE (1989) K-Ar dating of illite in hydrocarbon reservoirs. Clay minerals, 24: 215-231

Hower J, Eslinger EV, Hower EM, Perry EA (1976) Mechanism of burial metamorphism of argillaceous sediment: 1. Mineralogical and chemical evidence. Geol Soc Amer Bull, 87: 725-737

Kralik M (1984) Effects of cation-exchange treatment and acid leaching on the Rb-Sr system of illite from Fithian. Illinois. Geochim Cosmochim Acta, 48: 527-533

Liewig N, Clauer N, Sommer F (1987) Rb-Sr and K-Ar Dating of Clay Diagenesis in Jurassic Sandstone Oil Reservoir, North Sea. Am Assoc Petrol Geol Bull, 71:1467-1474

Odin GS, Matter A (1981) de glauconiarum origine. Sedimentology, 28: 611-641

Odin GS, Dodson MH (1982) Zero isotopic age of glauconies. In: GS Odin (ed) Numerical Dating in Stratigraphy. John Wiley and Sons Publ, Chichester: 277-305

Odin GS, Hunziker H (1982) Radiometric dating of the Albian-Cenomanian boundary. In: GS Odin (edit.) Numerical Dating in Stratigraphy. John Wiley and Sons Publ, Chichester: 537-556

Odin GS, Fullagar PD (1988) in Green Marine Clays, Development in Sedimentology 45, Elsevier: 295-332

Odin GS (1988) Green Marine Clays. Developments in Sedimentology 45, Elsevier

Savin SM, Epstein S (1970) The oxygen and hydrogen isotope geochemistry of ocean sediments and shales. Geochim Cosmochim Acta, 34: 43-63

Savin SM, Lee M (1988) Isotopic studies of phyllo-silikates. In: Hydrous phyllosilikates (excl. of micas), Bailey S.W. (Ed.), Reviews in Mineralogy, Min Soc Amer, 19: 189-223

Stille P, Clauer N (1994) The process of Glauconitization. Chemical and isotopic evidence. Contr Mineral Petrol, 117: 253-262

Yeh HW, Savin SM (1976) The extent of oxygen isotope exchange between clay minerals and seawater. Geochim Cosmochim Acta, 40: 743-748

Yeh HW, Savin SM (1977) Mechanism of burial metamorphism of argillaceous sediments: 3. O-isotope evidence. Geol Soc Amer Bull., 88: 1321-1330

Yeh HW, Epstein S (1978) Hydrogen isotope exchange between clay minerals and seawater. Geochim Cosmochim Acta, 42: 140-143

Yeh HW (1980) D/H ratios and late stage dehydration of shales during burial. Geochim Cosmochim Acta, 44: 341-352

Yeh HW, Eslinger EV (1986) Oxygen isotopes and the extent of diagenesis of clay minerals during sedimentation and burial in the sea. Clays and Clay Minerals, 34: 403-406

7 The Sm-Nd Isotope System in Detrital and Authigenic Argillaceous Sediments

As was shown in the previous chapter, classic isotopic methods such as K-Ar and Rb-Sr allow us to date authigenic mineral components found in detrital sediments and sedimentary rocks. We can date diagenetic, hydrothermal or metamorphic events provided we can separate the associated authigenic mineral components cleanly. The presence of detrital components, carrying various isotopic compositions, can however render ages geologically irrelevant. Isotopic homogenization between migrating fluids and authigenic mineral components at high temperatures can also lead to a reopening of the isotopic systems of authigenic clay minerals as they adapt to new physical and chemical conditions. This is especially relevant for the Rb-Sr and K-Ar isotopic systems and has the consequence that mineral phases of more ancient systems frequently remain open. That Rb-Sr and K-Ar systems hardly ever permit the correct absolute age dating of early diagenesis in detrital sediments was demonstrated by the first systematic Sm-Nd isotopic studies on sediments (Stille and Clauer 1986; Schaltegger et al. 1993). The Sm-Nd age can also be thermally and chemically reset after deposition and diagenesis (Toulkeridis et al. 1990; 1994). However, in these studies and others it was shown that the rare earth elements, including the Sm-Nd isotopic system, are far more robust against fluid migration in sedimentary rocks than either the Rb-Sr or the K-Ar systems. This can be understood better using the example of the clay-rich sedimentary rocks of the Proterozoic Gunflint Banded Iron Formation (BIF).

7.1 The First Attempts to Date Diagenesis Using the Sm-Nd Isotope Method

The very first Sm-Nd isotopic study, which dealt systematically with the age and genesis of clay-rich sedimentary rocks was reported by Stille and Clauer (1986). The authors investigated clay minerals deposited in the iron-rich horizons of the

Proterozoic Gunflint "Banded-Iron Formation" in Canada using not only isotopic methods but also various geochemical and mineralogic approaches.

The formation stretches from the north coast of Lake Superior over 180 km westwards along the border between Minnesota und Ontario. This sedimentary succession is not only famous due to its rich iron deposits but also because of the appearance of ancient, primitive organisms. The microflora that has been found here has provided information about the early chemical evolution of the atmosphere and the nature and origins of the earliest organisms. Deposition of the sediments that alternate with the iron ores took place in a slowly subsiding shallow water basin, perhaps comparable to the depositional environment of evaporites and phosphorites in the Phanerozoic. The basin formed around 2100 million years ago. The exact time of deposition and formation is not yet known and is the focal point of several discussions. Isotopic age data from the literature are contradictory. Already in 1966, Peterman suggested that the Rb-Sr and K-Ar systems had not remained closed for some of these sedimentary rocks.

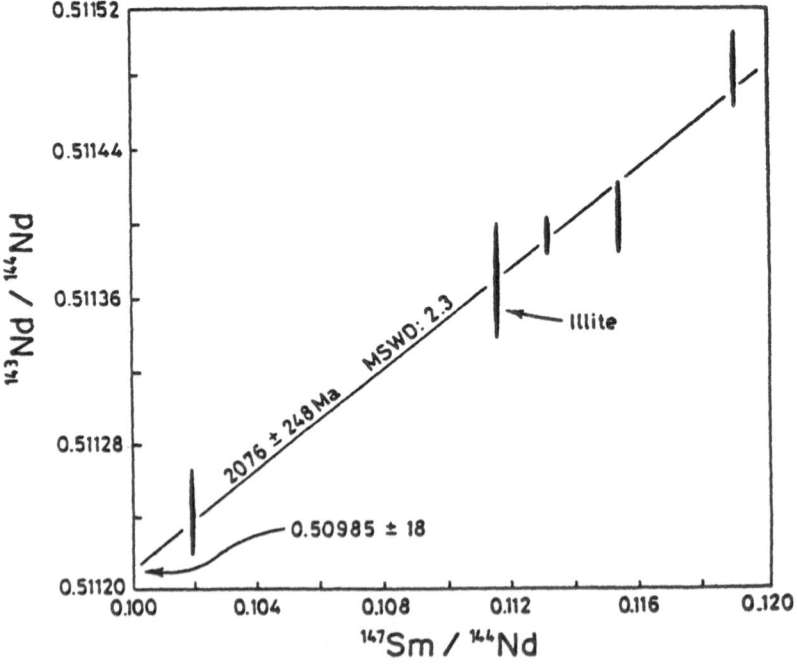

Fig. 7.1. Sm-Nd isochron diagram for clay minerals (chlorite and illite) from clay-rich rocks of the 'Banded Iron Formation' in Canada. The gradient of the line corresponds to an age of 2.08 +/- 0.25 Ga and is interpreted as the age of formation (timing of diagenesis) of the clay minerals. (Stille and Clauer 1986)

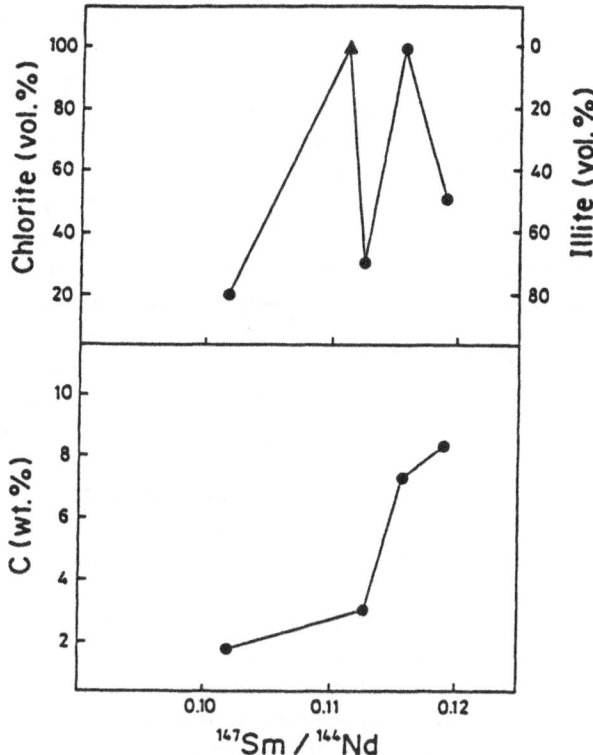

Fig. 7.2. Chlorite content and organic carbon concentration from the Gunflint-clays as a function of their Sm/Nd ratios. (Stille and Clauer 1986).

Stille and Clauer (1986) selected clay minerals for their Sm-Nd age dating, mostly illites and chlorites. The illite crystallinities and the common presence of pure chlorite implied a primary, diagenetic origin for these mineral assemblages. The clays contained up to 9% organic carbon. Investigations showed that the organic matter consisted largely of pristane, phytane and paraffin. Therefore, these were characteristic chemical sediments, with oil, primary diagenetic minerals and extremely high contents of biogenic carbon. Clay minerals of this type could have possessed Sr and Nd isotopic compositions at the time of their formation that were fully equilibrated with seawater. Samples for isotopic analysis were chosen with various chlorite/illite ratios and organic carbon contents in order to obtain the largest possible variation in Rb/Sr and Sm/Nd ratios.

The K-Ar ages of these samples were scattered between 1.57 and 1.38 Ga. The gradient of the Rb-Sr isochron indicated an age of 1.46 Ga. Similar variations in the K-Ar and Rb-Sr ages were obtained from stratigraphically neighbouring, very brown, oxidised clays (Peterman 1966). Peterman assumed that the sediments had been altered by hydrothermal weathering some 1.46 Ga ago.

In this study, Stille and Clauer were able to observe for the first time that the behaviour of the Sm-Nd isotopic system can be decoupled from that of the Rb-Sr and K-Ar isotopic systems. Four whole rock fractions, which were shown on the basis of X-ray diffraction to consist of authigenic clay minerals (chlorite, illite), and a pure chlorite fraction, defined an isochron on the Sm-Nd diagram giving an age of 2.08 ± 0.25 Ga (Fig. 7.1). The relatively high error results from the quite small variation in the Sm/Nd ratios of these fractions. The Sm/Nd ratios appeared to be more dependent on the organic carbon content than the illite/chlorite ratios as the REE tend to form complexes with organic matter that have higher Sm/Nd ratios than clay minerals. The incorporation of different amounts of organic carbon into the clays led to a positive correlation between organic carbon content and Sm/Nd ratios (Fig. 7.2). The influence of organic carbon on the mineralogic and chemical development of detrital sediments was already discussed in Sect. 6.1. and will be referred to again in Sect. 7.2. in the light of the Sm-Nd isotopic system.

Geologic constraints confirm this older age of deposition, i.e. around 2.08 Ga. However doubts still creep in concerning the absolute correctness of this age interpretation. For example, the possible presence of detritus leaves questions unanswered especially as we know that their presence can remain undetected using standard methods such as X-ray diffraction. Undetected minerals, such as apatite, with high REE concentrations can alter the Sm-Nd isochron age significantly (see Sect. 7.3). Therefore, in order to date diagenetic processes using the Sm-Nd isochron system, other approaches must be applied that allow us to identify small quantities of REE-rich phases and to analyze them both chemically and isotopically. Also the question of to what extent Nd isotopic homogenization can be achieved between authigenic and detrital mineral phases could not be answered for the Gunflint Formation. Nevertheless, this study was of great importance for the work that was to follow, as the possibility of dating diagenetic processes using the Sm-Nd isotopic system had been clearly demonstrated. Also the influence of organic matter on Nd isotopic homogenization during diagenesis could be recognized. In the following sections, we will discuss in more detail the influence of both organic carbon and REE-rich accessory minerals on the Sm-Nd isotopic systems of clay-rich sediments and sedimentary rocks.

7.2 The Process of Nd Isotopic Homogenization in Bituminous Shales

Isotope studies of Proterozoic black shales from Gabon (Africa) permitted a full description of the process of Nd isotopic equilibration between authigenic mineral phases, fluid phases and oil (Bros et al. 1992; Stille et al. 1993). These sediments derived from 1000 meter thick sedimentary piles of the "Francevillian", which were deposited directly on Archean crystalline basement. The fine-grained pelites and black shales alternate with sandstone, dolomite and ignimbrite tuff horizons.

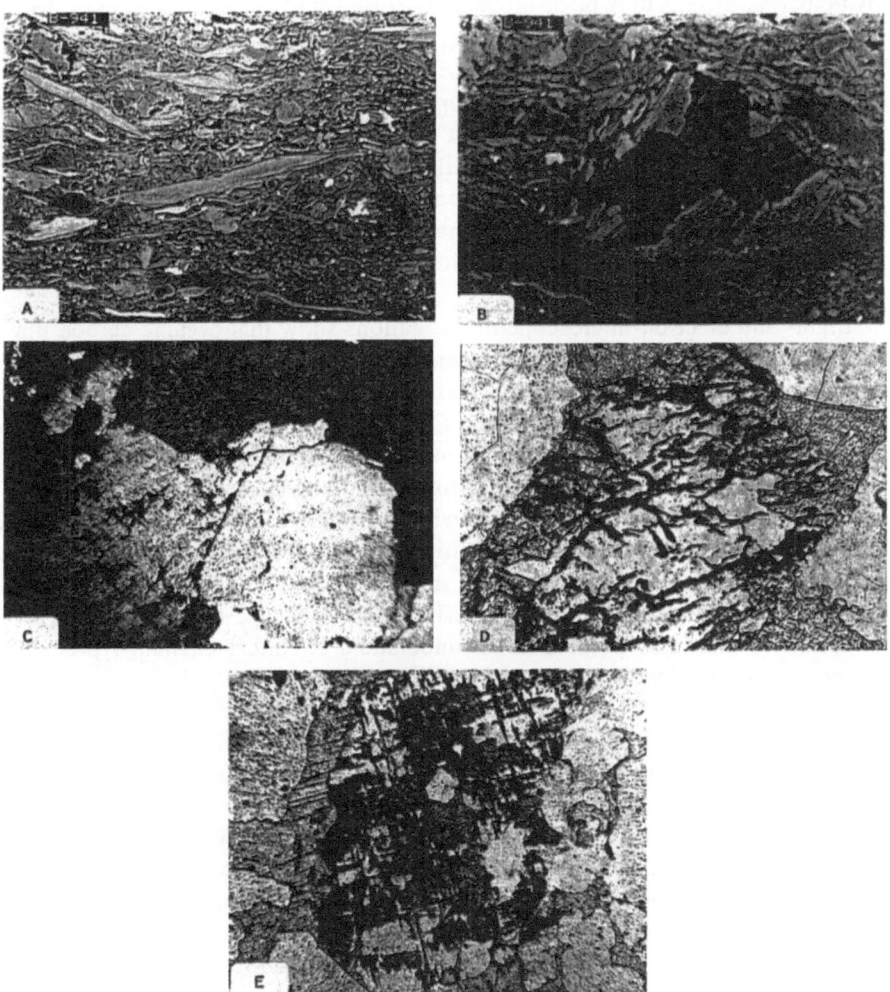

Fig. 7.3. Mineralogy of the bituminous clay-shales from Gabon. A) Detrital, aligned biotites. B) Authigenic clay minerals (illite, chlorite) in the clay matrix. C) Silicified and decomposed alkali feldspars. D) Organic matter concentrated along the cleavage planes of the alkali feldspars. E) Detrital alkali feldspar is replaced by the growth of fresh illite and chlorite. (Stille et al. 1993)

The shale layers are also enriched in organic carbon, which can make up to 15% of the whole rock weight. This series is also well known for the richness of exploitable uranium deposits there, in particular Oklo where natural nuclear reactions took place 2 Ga ago. On the basis of isotopic investigations (K-Ar, Rb-Sr) we know that, as with the "Gunflint Banded Iron Formation", the last hydrothermal events took place around 2.0 to 1.8 Ga ago (Bonhomme et al. 1982). The sediments themselves must be older than 2.0 Ga. The most important detrital mineral components that occur in the fossil-oil (i.e., solid bitumen)-rich black shales are quartz and large, up to 500 μm biotites, which are aligned parallel to the primary sedimentary bedding and 'float' in a much finer rock matrix (Fig. 7.3A). Silicate minerals such as chlorite and illite formed during diagenesis. They are found in the matrix of the rock and reach grain-sizes of just 0.5 μm, i.e. 1000 times smaller than the large detrital biotites (Fig. 7.3B). Mineralogic investigations show that these chlorites and illites derived from detrital K-feldspars (microclines) (Fig. 7.3E).

Light- and electron-microscope investigations show that organic carbon plays an important role in feldspar replacement and is also involved in the new formation of chlorite and illite (Fig. 7.3D). The organic matter lies on top of the crystal faces and in the surface cracks of the often strongly corroded and silicified feldspars, thus mimicking the original microcline twinning crystal structure. The influence of organic hydrocarbons on long-term diagenetic processes and on the new formation of illites and chlorites will be discussed more fully later on. The following discussion of the results of various Nd isotopic analyses will allow us to recognize that the Nd isotopic system in these rocks provides a very important tool to the understanding of their genesis. Coupling of the Nd isotope data with mineralogic data also permits us to understand chemical exchange processes which have taken place between the silicate phases and oil.

Four black shales from various depths in one bore hole, and pure bitumen, which migrated out of the clay shales and ended up in the joints and cracks of the overlying sandstone horizon, were used in these isotopic investigations. Various grain-size fractions: 2-0.8, 0.8-0.4, 0.4-0.2 and <0.2 μm were extracted and concentrated from two of these shales. Concentrating these various clay fractions made it possible to separate the diagenetically produced clay minerals with grain size <0.5 μm from the detrital clay minerals with grain sizes up to 500 μm.

Leaching experiments were carried out using 1N HCl acid on the smallest fractions (<0.4 μm). Clauer et al. (1993) assumed that the leachates were isotopically and geochemically identical to the fluid phase, with which the clay minerals last experienced exchange. If the clay minerals crystallized within this fluid medium then they would have existed in isotopic equilibrium with this fluid phase. Leachates would thus produce leachate-residue internal isochrons that define the age of crystallization. The leachate represents the diagenetic fluid phase. Similar considerations can be brought to bear on the Sm-Nd isotopic system. Interpretations may be problematic due to the possible existence of small

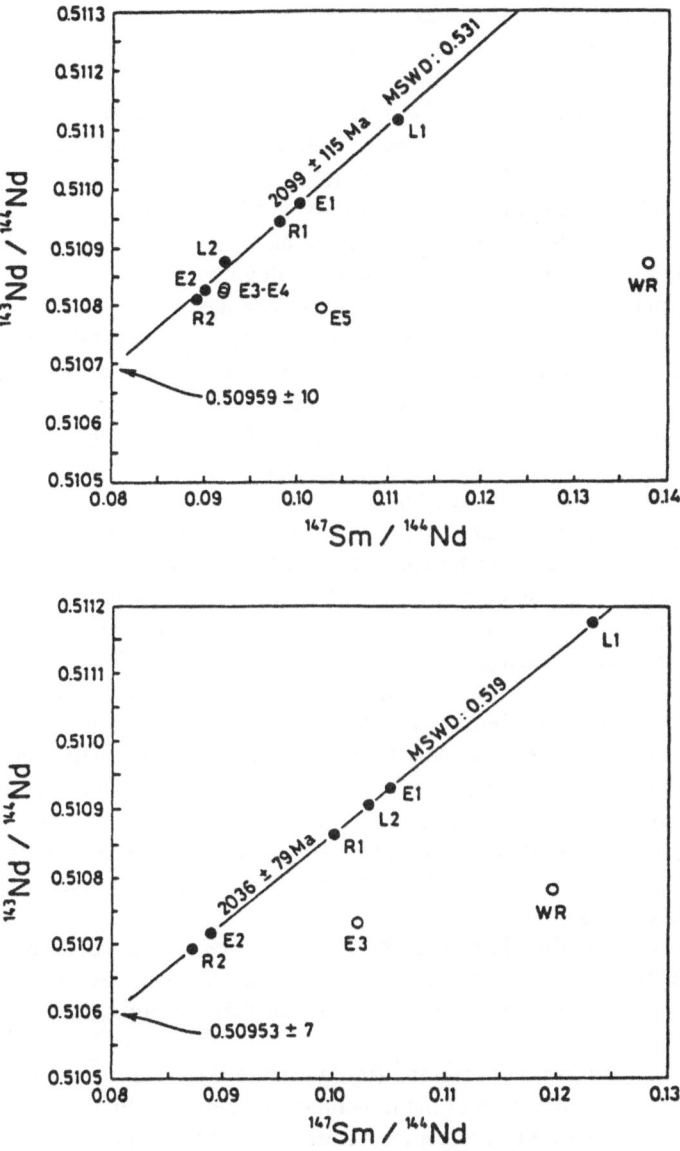

Fig. 7.4. Sm-Nd isochron diagram for two bituminous clay-shale samples from Gabon. WR: whole rock; E1: clay fraction <0.2 μm; E2: Clay fraction <0.4 μm. L1, L2, R1, R2: Leachate-residue pairs for E1 and E2 respectively. E3-E5: clay fractions >0.4 μm. Only the clay fractions E1, E2 and the corresponding leachate-residue pairs define isochrons whose ages are identical within the margin of error. The initial $^{143}Nd/^{144}Nd$ ratios of these isochrons are also identical at 0.50959 +/- 10 and 0.50953 +/- 7 (MSWD = mean square of weighted deviates). (Bros et al. 1992)

amounts of partially or totally acid-soluble accessory phases, such as apatite or iron hydroxides, which are likely to be strongly enriched in the REE. Thus, the isotopic system of the authigenic minerals will be disturbed should this apatite be of detrital origin. Precise dating of diagenesis would not be possible in this case. However, if these phases were cogenetic with the authigenic clay minerals, they will also have incorporated the same, initial isotopic composition. The possible presence of phosphatic minerals is always a factor to be considered when interpreting the Nd isotopic ratios of clay minerals in sedimentary rocks. Chemical analyses (not less reliable petrographic or mineralogic analyses) demonstrated that there was no phosphate in the leachates of these samples.

Not only the untreated samples, but also the leachates and the residues were investigated for their Sr and Nd isotopic compositions. The granulometric fractions, as well as the leachates and residues of two samples yielded closely parallel straight lines on the isochron diagrams. The gradients of these straight lines are nearly identical and yielded identical ages within error of 2036 ± 79 und 2099 ± 115 Ma (Fig. 7.4). The leachates show systematically higher Sm/Nd ratios than the corresponding residues (R) and the untreated clay fractions (E). According to the law of mass-balance, the residues, leachates and the untreated clay-fractions should all lie on a straight line on the isochron diagram.

Provided the clay minerals were in isotopic equilibrium with the coexisting diagenetic fluid phase during their crystallization, the Sm/Nd isotopic ratios of this fluid phase should correspond to those of the leachates. The resulting parallel isochrons and their gradients should correspond to the time of isotopic equilibrium, crystallization and thus of diagenesis. The initial ratios would thus reflect the Nd isotopic composition of the environment in which the clay minerals crystallized. If, however, the clay mineral fractions contained inherited detrital components, then the gradients will represent a geologically irrelevant age and the straight lines will correspond to a mixture between detritus and later diagenetic phases (see Sect. 6.3).

Interpretation of the data shown here requires a careful mineralogic, morphologic and chemical characterization of the investigated material. The following observations allow us to consider that these straight lines are indded geologically relevant.

1) X-ray diffraction investigations (to determine the illite crystallinity index and to analyze the composition of the mineral components qualitatively), chemical analyses (in order to investigate the mineral assemblages quantitatively) and morphologic observations using the electron microscope, show us that the smallest clay fractions are enriched in 1M illite (authigenic). A significant amount of detrital components could only be found in coarser fractions and in the whole rock. Fig. 7.5 shows the well crystallized 1M illite from the smallest clay fractions compared with the detrital illite from the coarser fractions.

2) Two different clay fractions, the associated residues, and leachates of two whole rocks, which were collected at two different localities, define straight lines

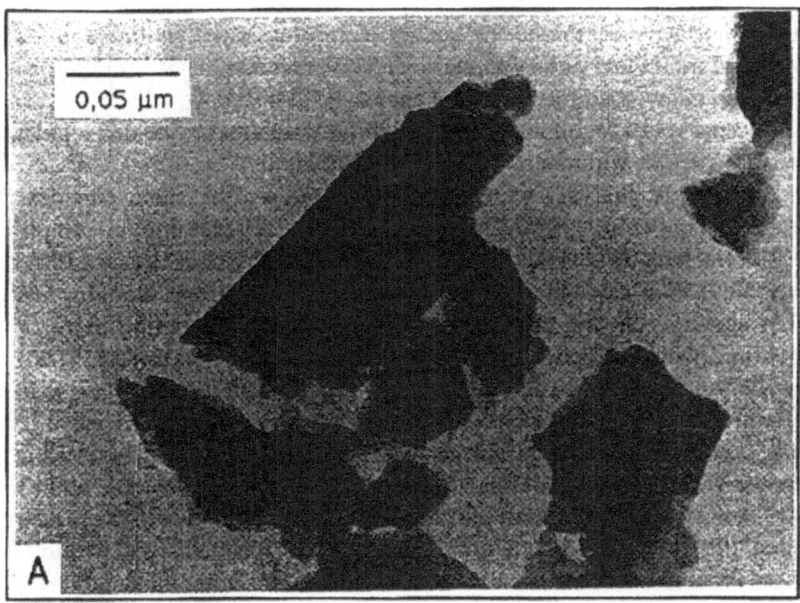

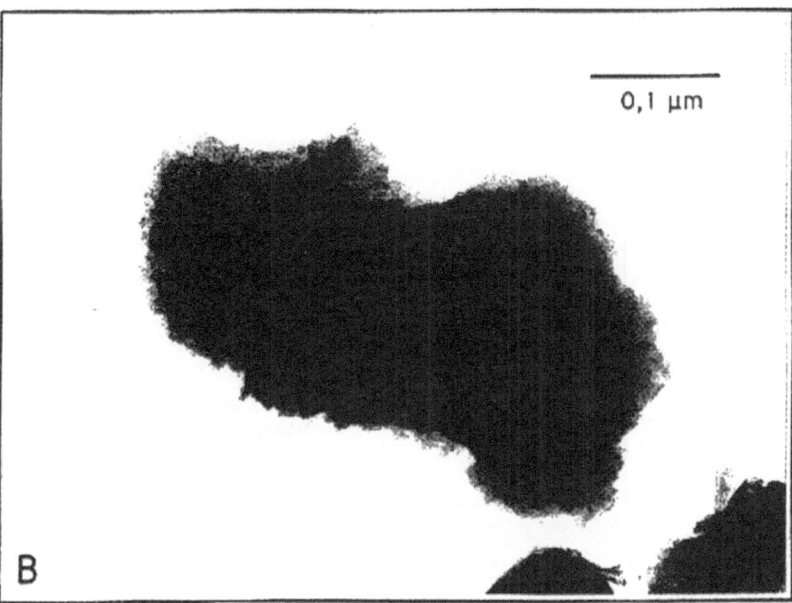

Fig. 7.5. Electron microscope photographs of clay fractions from the bituminous black shales. A) <0.2 μm clay fractions; B) 0.8-2.0 μm clay fraction. (Bros et al. 1992)

on the isochron diagram with a precision that might more normally be found in the comparatively simple system of a magmatic rock. The presence of inherited detritus would have decreased the quality of resolution of this line and led to a scattering of individual points on the line. The gradients of the straight lines are identical and allow, therefore, identical ages to be obtained.

3) Volcanic rocks found interbedded within the sediments yielded an identical Rb-Sr isochron age, within error, of 2140 ± 140 Ma.

Pb-Pb isotopic analyses were also carried out on the same clay fractions (Gauthier-Lafaye et al. 1996). The Pb isotope data suggest a more complicated story. The leachate-residue pairs of the <0.4 µm clay fractions define a Pb-Pb age slightly older than that obtained from the Sm-Nd isochrons. Looking in more detail it appears that the different leachate-residue pairs define more or less parallel lines with a steeper slope than the general regression line (Fig. 7.6). The subparallel lines defined by the different leachate-residue pairs suggest that each clay fraction has its own specific inherited Pb component. Therefore, if the isotopic composition of the leachate was that of the last diagenetic fluid that exchanged with the clay minerals, it never reached total Pb isotopic equilibrium with the non-exchangeable Pb of the clay minerals probably due to the presence of inherited Pb. In the Pb-Pb diagram of Fig. 7.7., leachates and residues are regressed separately and define alignments with slightly different slopes.

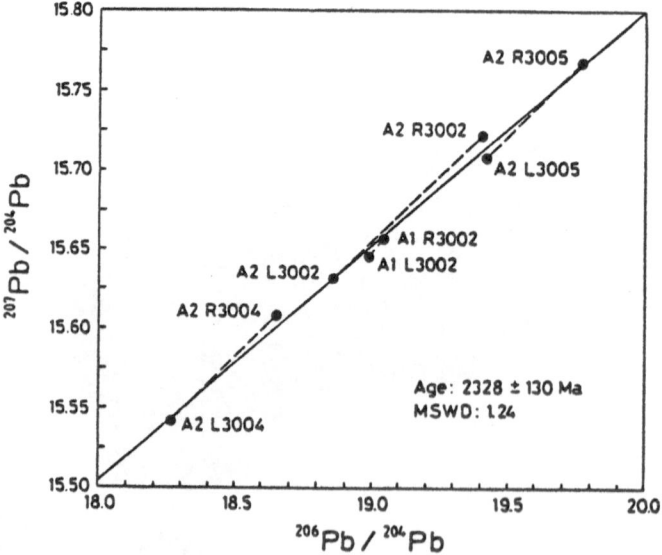

Fig. 7.6. $^{207}Pb/^{204}Pb$ and $^{206}Pb/^{204}Pb$ ratios of leachates (L) and clay residues (R). A1:<0.2 µm; A2: 0.2-0.4 µm. Full line: general regression line with all data points. The leachate-residue tie lines (dashed line) for individual samples have greater slopes. (Gauthier-Lafaye et al. 1996)

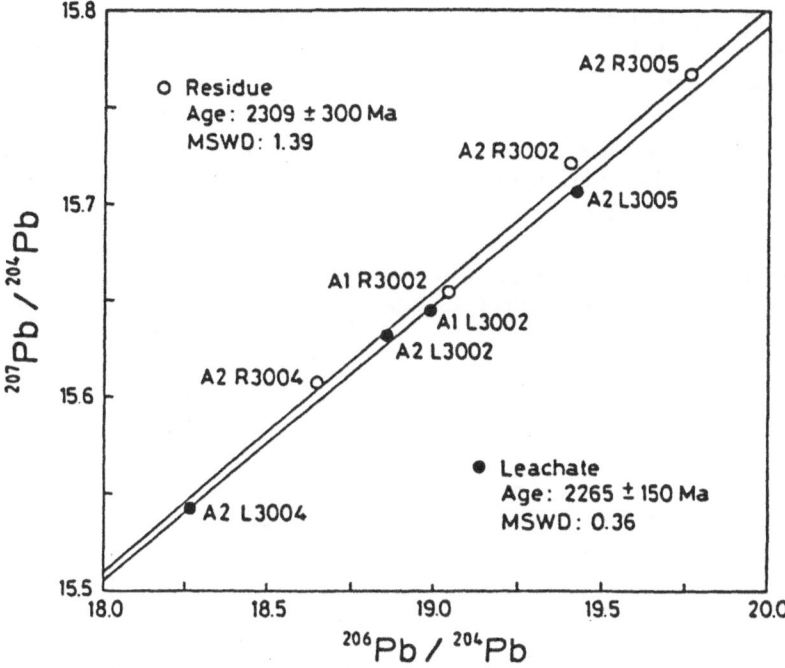

Fig. 7.7. ^{207}Pb/^{204}Pb vs. ^{206}Pb/^{204}Pb diagram showing the regression line for leachates and residues. (Gauthier Lafaye et al. 1996)

The age of 2.26 ± 0.15 Ga obtained for the leachates is identical to the Sm-Nd age. The age obtained for the residues is identical to that of the leachates within analytic errors. However, the slightly steeper slope of the residue regression line and the much stronger scatter of the data points as reflected by the much higher error in age probably point to the presence of some inherited Pb. The evidence of inheritance in the residues suggests that most probably only the leachates reached isotopic equilibrium and yield reliable Nd and Pb ages.

Let us get back to the Sm-Nd isochrons. The precision of both isochrons is boosted by the strikingly large variation in the Sm-Nd ratios. Which mechanism could be responsible for such a fractionation? The leachates show higher Sm/Nd ratios than the residues. The Sm/Nd ratios of the residues themselves are strongly related to the grain-size of the clay-fraction: the finer the fraction, the higher the Sm/Nd ratio of the leachate and also of the untreated clay-fraction. As already mentioned, it can be assumed that the leachates represent the fluid phase in which the clay minerals crystallized. This implies that the Sm/Nd ratios in the fluid phase are controlled by the precipitation and crystallization of increasingly smaller authigenic clay minerals and allows us to further assume that:

1) The authigenic clay minerals are enriched in Nd over Sm and, 2) the crystallization of Nd-rich clay minerals led to the successive depletion of Nd in the fluid phase and also, therefore, to a consistent rise in the Sm/Nd ratios of the later diagenetic fluids. The Sm-Nd fractionation would thus be comparable in

nature to fractional crystallization of a granitic melt. A further component, i.e. the oil which migrated into these rocks, may have had an additional influence on this fractionation process. This is demonstrated by the Sm-Nd isotopic analyses on oil-rich clays and oil-bitumens which solidified in the higher porosity of the overlying sandstone. The negative correlation between the Sm/Nd ratios and the organic carbon content (Fig. 7.8C) demonstrates that the whole rock samples richest in oil have the lowest Sm/Nd ratios. These samples lie close to the lower ends of both isochrons (sample 3005 in Fig. 7.9). It is also conceivable that the oil-rich clays existed in isotopic equilibrium with the diagenetic fluid phases and the clay minerals. The probability of Nd isotopic equilibrium was already discussed by Manning et al. (1991).

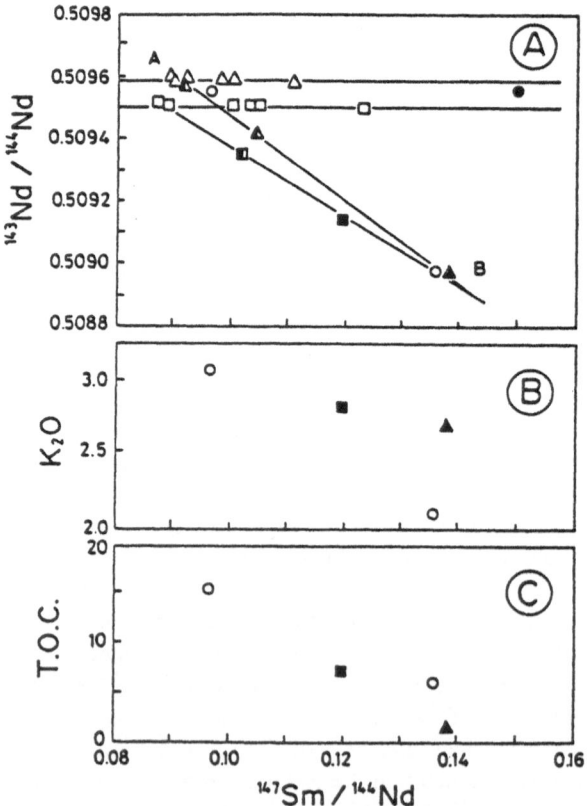

Fig. 7.8. (A) Nd isochron diagram. The Nd isotopic ratios of the clay fractions and their respective whole rocks have been corrected to a diagenetic age of 2.06 Ga. Open triangles and squares: <0.4 μm clay fractions. Filled circles: bitumen. Half filled circles and squares: >0.4 μm clay fractions. Filled squares and triangles: corresponding whole rocks for the clay fractions. (B) and (C) show the dependence of the Sm/Nd ratio on the K_2O and the organic carbon contents (TOC). (Bros et al. 1992)

The oil which migrated at that time and has today become our solid bitumen shows strongly increased $^{143}Nd/^{144}Nd$ and Sm/Nd ratios and lies similarly on the extension of both isochrons (Fig. 7.9). Its isotopic characteristics allow us to assume that isotopic equilibrium was achieved between the oil, the diagenetic fluids and the clay minerals. They also show that the oil-bearing black shale could represent the parent rock of the migrated oil. The strikingly high Sm/Nd ratio of the bitumens can be the result of the maturation process that these oils experienced during diagenesis.

Manning et al. (1991) were able to demonstrate that the Sm/Nd ratios increase with increasing degree of maturity. The authors assume that the heavy REE are preferred over the lighter REE by organo-metallic complexes during formation. Thus Sm is more likely to be incorporated into the complexes than Nd, which has the consequence that migration and progressive maturation of the oil will tend to increase the Sm/Nd ratio. If close isotopic exchange has taken place between the migrating oil and the other fluid phases, then the formation of the organo-metallic complexes will have played an important role in the chemical evolution of the fluid phases and the development of the Sm/Nd ratio of the authigenic clay minerals. Apparently, isotopic equilibrium can be achieved between clay minerals and oil, whereby the formation of organo-metallic complexes and crystallization of chlorite and illite can sufficiently fractionate the Sm/Nd ratios of the diagenetic fluid phase to allow dating of the diagenetic crystallization of the clay minerals.

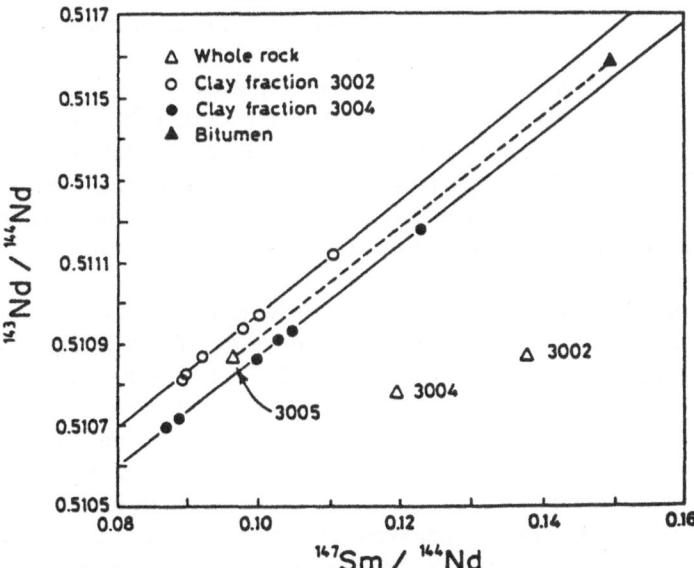

Fig. 7.9. This Sm-Nd isochron diagram shows the close relationship between bitumen and whole rock with the highest contents of organic carbon (<15%; sample 3005) and between bitumen and the clay fractions (<0.4 μm) of samples 3002 and 3004. (Bros et al. 1992)

However, this only goes some way to answering all the questions related to chemical exchange between the newly formed clay minerals, detritus, organic hydrocarbons and the fluid diagenetic phases. How, for example, would this fluid phase crystallize, in which the clay minerals are formed? Where did it come from? Has there been large exchange of rock volume? Has the fluid travelled far? Did chemical exchange occur between the fluid phase and the oil?

Some answers to these questions can come from investigations of the isotopic characteristics of the coarser fractions and the whole rock at the time of sedimentation and diagenesis as they are predominantly carriers of the detrital components. In the isochron diagram of Fig. 7.8A, the Nd isotopic data for the clay fraction and the whole rock have been corrected for a diagenetic and sedimentary age of 2.06 Ga. This is important to remember when considering these data!

The finest clay fractions must in this case, according to the definition of the isochron, possess identical initial Nd isotopic ratios to the bitumen and thus come to rest on horizontal lines. The coarser clay fractions and the corresponding whole rocks define, instead, two straight lines with negative gradients. They cut the horizontal lines on the low Sm/Nd ratio side at a point defined by the coarsest clay fractions (<0.4 µm), which also help to define the corresponding isochrons.

The $^{147}Sm/^{144}Nd$ ratios of these junctions are low and vary between 0.085 and 0.09. The Sm/Nd ratios of the whole rocks and the coarser clay fractions rise with decreasing potassium and organic carbon content (Fig. 7.8 B, C). The sample with the highest potassium and organic carbon contents lies between two horizontal lines, as does the bitumen, and yields an identical Nd isotopic composition to that of the finest clay fractions of the other two whole rocks. This observation allows us to make some conclusions regarding the origins of the detritus as well as shedding light on the diagenetic evolution of the oil-bearing clay shale.

Now to the origin of the detritus: The negative correlation between the $^{143}Nd/^{144}Nd$ and Sm/Nd isotopic ratios permits us to assume that these sediments contain two isotopically different and important detrital components A and B that were mixed together in different proportions at the time of sedimentation of this rock.

Component A shows a Sm/Nd ratio which is far lower than the mean of all clay shales or crustal rocks. Scarcely any mineral other than alkali feldspar possesses such low ratios. This confirms Norm-composition calculations, which show that the normative contents of alkali feldspar increase with decreasing Sm/Nd ratios and increasing $^{143}Nd/^{144}Nd$ isotopic ratios. The whole rock, which lies closest to the junction of the straight lines contains the highest potassium and normative alkali feldspar contents. The mixing lines AB reflect therefore mixtures of a potassium-rich alkali feldspar component A with another detrital component B, which must have far lower $^{143}Nd/^{144}Nd$ and higher Sm/Nd isotopic ratios. With the help of crustal residence time calculations, the origins of the rare earth elements and the detritus could be established (Stille et al. 1993).

Let us return to the diagenetic development of these rocks. It was already mentioned that the formation of the new clay minerals and organo-metallic complexes can have the consequence that the Sm/Nd ratios of the diagenetic fluid phase are fractionated. The formation of illite seems to be related to the presence and abundance of potassium-rich mineral phases and organic carbon in the clay shales.

Whole rocks and coarse clay fractions approach the horizontal equilibrium lines of the newly formed mineral phases along the mixing line AB with decreasing Sm/Nd ratios and increasing K and Corg contents (Fig. 7.8A). Provided the ^{147}Sm/^{144}Nd ratios reflect the previous presence of alkali feldspar in these rocks at the point of junction with the horizontal lines, then it is probable that the newly formed chlorite, illite, bitumen and the whole rock with the highest K and TOC contents all record the Nd isotopic composition of this feldspar on the basis of their identical Nd isotopic ratios. The close genetic relationship between chlorite, illite and K-feldspar has already been mentioned (see Fig. 7.3C).

The diagenetic mineral reactions discussed in Chap. 6.1 show that organic acids play an important role during diagenesis. Changes in the partial pressure of CO_2 and the concentration of organic acids control dissolution and replacement reactions and the secondary porosity of a rock. Dissolution of carbonates and aluminium silicates takes place predominantly under acidic conditions, whereby etching and oxidation will lead to the release of organic carbon (CH_2O).

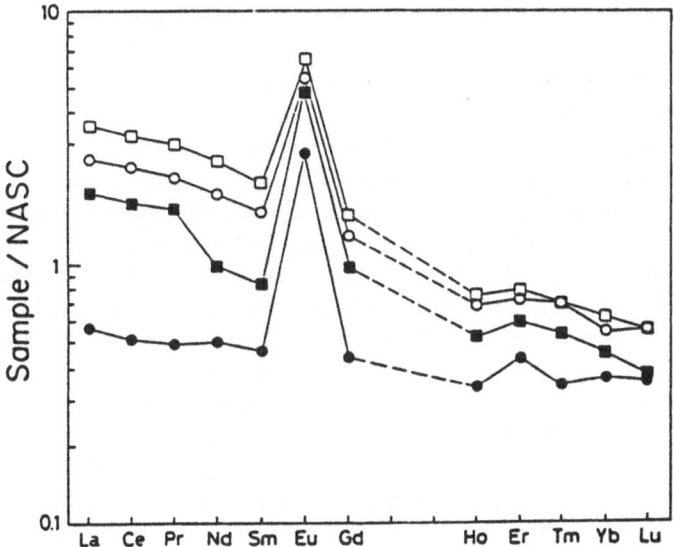

Fig. 7.10. REE distribution patterns of different clay size fractions. The <0.2 μm fractions show lowest REE concentrations and the coarsest clay size fractions (0.8-2 μm) show the highest REE concentrations. (Bros 1993)

The oxidation of organic carbon and reduction of sulphate will lead to the production of CO_2 and H_2O and the formation of carbonic acid. The presence of relatively large quantities of pyrite in the investigated samples suggest that sufficient sulfur atoms were indeed available in order for such reactions to take place. The carbonic acid makes further dissolution of alkali feldspar and the new formation of chlorite and illite possible. It does not only help in dissolution reactions but also in the subsequent silicification of the alkali feldspar (Fig. 7.3C; see also Chap. 6.1). The REE patterns of the clay fractions with the pronounced Eu (europium) anomaly are typical of feldspar and confirm the close relationship between illite and feldspar (Fig. 7.10). As feldspars, oil and sulfides coexisted, we can assume that the process of feldspar dissolution and illite-chlorite crystallization took place **in situ** and without the migration of fluid phases across large distances. Isotopic exchange with other silicate phases from neighbouring rock associations can therefore be excluded (Stille et al. 1993).

As both the organic carbon and the sulfides are depleted in REE (their concentrations are in the ppt-range), it can be assumed that the Nd isotopic composition of the newly formed clay minerals were controlled by the replacement of the feldspars only. Provided these detrital feldspars possessed nearly identical, initial isotopic ratios, this means to say that they were derived from the same source area, then the subsequently formed illites will have incorporated the same isotopic composition. These clay minerals define isochrons, whose gradients correspond to the age of crystallization and whose initial isotopic composition would reflect that of the detrital feldspars at the time of diagenesis. This isotopic composition corresponds to that of the fluid phase in which illite and chlorite formed. Nd isotopic homogenization of this sort is likely to be important on a small-scale (hand specimen scale) but unlikely to be important in more sizeable rock successions.

The oil coexisted in isotopic equilibrium with the diagenetic fluids (aqueous formation waters). The mechanism that led to isotopic equilibrium is difficult to understand as oil and aqueous solutions are not able to mix under normal conditions. It is conceivable that diffusion played an important role. However, it is more likely that the isotopic composition of the REE-poor hydrocarbons were overprinted by the REE rich, acidic, watery solutions, which had already taken part in the etching and replacement of the feldspars. This second phase may have coexisted within the oil without the necessity for mixing. In addition it might be possible that the originally REE-poor hydrocarbons mainly contain REE adsorbed from the diagenetic fluid phases. Under these conditions, identical isotopic compositions may arise in both the oil and in the authigenic, silicate mineral phases. If we can identify all possible parent rocks, we can compare the isotopic compositions of authigenic mineral phases in these rocks and the oil in order to determine the source of the oil.

Whole rocks can only reflect the isotopic composition of the oil if they consist exclusively of authigenic mineral phases, which formed under the conditions just described. The whole rock sample with the highest potassium and organic carbon contents, with the highest proportion of newly formed chlorite and illite and free of detectable detrital biotite may have existed in isotopic equilibrium with the migrated oil and also with the diagenetic fluid phases.

The Nd isotopic homogenization model postulated for the Francevillian series of Gabon was confirmed in a later project involving a bituminous, upper Carboniferous clay succession from the Weiach borehole in the northwest Switzerland Permian-Carboniferous 'trough' (Schaltegger et al. 1994). These rocks are particularly suited to the study of diagenetic isotopic equilibrium as not only authigenic clay mineral phases together with the bitumen but also alkali feldspar is present in abundance. The claystones experienced prolonged overprinting by post-Variscan hydrothermal events. Some of these hydrothermal pulses occurred about 180 Ma ago as determined by K-Ar investigations on illite fractions (Schaltegger et al., 1994).

The <0.4μm clay fractions comprise mainly authigenic illite. The phosphate contents of the clay mineral fractions and their acid leachates lie beneath the analytic detection limit and allows us to assume that the samples contain no phosphate. As with the Proterozoic black shales from Gabon, the REE patterns, in particular the striking Eu anomalies (Fig. 7.11), typical for feldspars, allow us to assume that the authigenic illites incorporated REE from the dissolution of alkali feldspars. The REE patterns demonstrate a close relationship between illite and alkali feldspar. Awwiller (1994) observed that leachate and residue phases of bulk mudrock und <0.5 μm clay fractions from the borehole of the Gulf Coast (Texas, USA) yielded an even larger positive Eu-anomaly with depth. Apparently, Eu was lost during the feldpar dissolution process and was subsequently incorporated into the newly forming illite/smectite and other authigenic phases.

The Nd isotope data for the Weiach samples confirm the close genetic relationship between alkali feldspar and clay minerals.

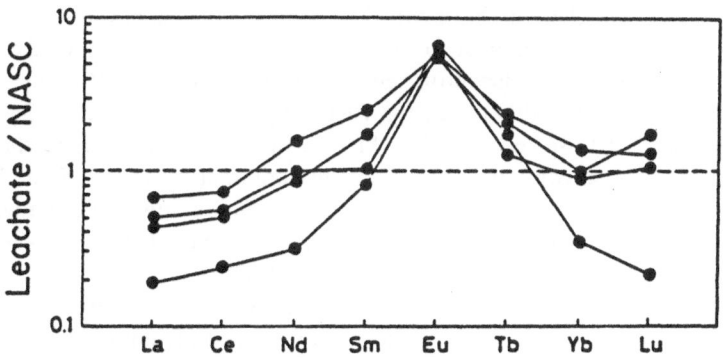

Fig. 7.11. Rare earth element patterns of leachates from Weiach sediments (normalized to NASC = North American shale).

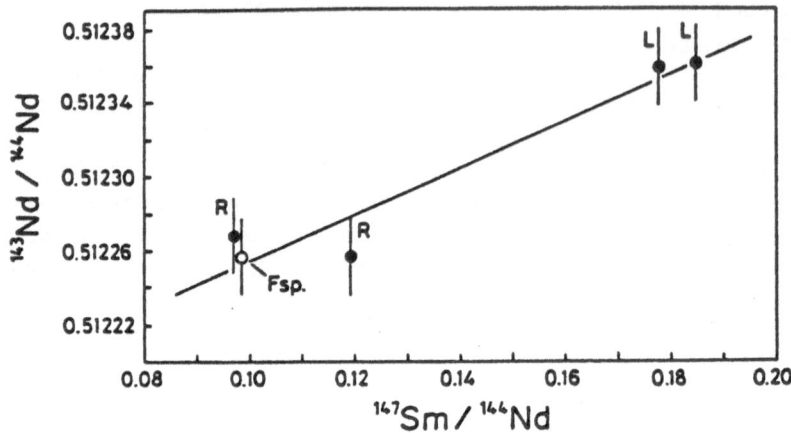

Fig. 7.12. Sm-Nd isochron diagram for bituminous clays (<0.4 μm) from Weiach (Switzerland). The data reflect isotopic equilibrium between illite and alkali feldspar (Fsp) and define an isochron whose gradient corresponds to an age of 192 +/- 29 Ma (MSWD = 1.43). L: leachates, R: residues.

Leachates and residues of the illites are found on the same straight line on the Sm/Nd isochron diagram as the coexisting alkali feldspars, whose gradient yields an age of 192 ± 29 Ma (Fig. 7.12). It can be assumed that this age reflects the timing of hydrothermal overprinting. In this case, as with the already discussed case of the Francevillian black shales (Stille et al. 1993), we can invoke a scenario involving the dissolution of K-feldspar, in situ crystallization of illite as well as Nd isotopic equilibrium between authigenic clay minerals. In this case, the illite adopted the Nd isotopic composition of the K-feldspar progenitor.

It is noticeable that for both the Oklo and the Weiach studies, the Rb-Sr isochron ages are not identical to the Sm-Nd ages. The clay minerals from the Oklo shales yield a Rb-Sr isochron age of 1875 ± 30 Ma years (Bonhomme et al. 1982) instead of around 2000 Ma (Fig. 7.4) using the Sm/Nd method. The Weiach samples yield a Rb-Sr age of 107 ± 11 Ma, instead of 192 Ma for the Sm/Nd method, see above. Similarly, the Gunflint sedimentary rocks (Sect. 7.1) also produced anomalously young Rb-Sr ages. How can we interpret these ages? Are they at all geologically relevant? Which mechanism can open the Rb-Sr isotopic system without opening the Sm-Nd isotopic system?

In the case of the Gunflint rocks, it can be shown on the basis of mineralogic, geochemical and isotopic investigations that the sedimentary rocks underwent alteration as a result of hydrothermal activity around 1500 million years ago, long after deposition and diagenesis (Peterman 1966). The Oklo sediments also show evidence of late carbonate diagenesis (Bonhomme et al. 1982). Rb-Sr, K-Ar and U-Th-Pb investigations on U bearing ore minerals allow us, in the case of Weiach, to establish that the sediments were hydrothermally overprinted several times sometime between 100 and 180 Ma. The leachates and residues of the Weiach samples describe a straight line on the Sr mixing diagram and allow us to suppose that the straight line relationship observed is likely to represent an "errorchron", i.e. no age can be resolved from its gradient. The chemical analyses reveal that the leachates of the clay minerals are strongly enriched in calcium and strontium (Ca: ~20 wt.%; Sr: 3000-10000 ppm). It is therefore probable that the last phase which enjoyed chemical exchange with the clay minerals represents a late carbonate diagenesis in the case of Weiach too.

How can the influence of such diagenetic alteration reveal itself in the Rb-Sr isotopic system? Let us assume that the Sr isotopic composition of the carbonate and strontium-rich fluid phase is lower than the Sr isotopic composition of the clay minerals with which it underwent exchange (Fig. 7.13). After exchange and mixing with the clay minerals, the carbonate phase will serve to pull each sample point of the isochron along the mixing line (C-1,C-2,C-3) away from the isochron and towards carbonate strontium. If all the samples have been mixed with the carbonate phase to the same extent, the samples will come to lie on a straight line again.

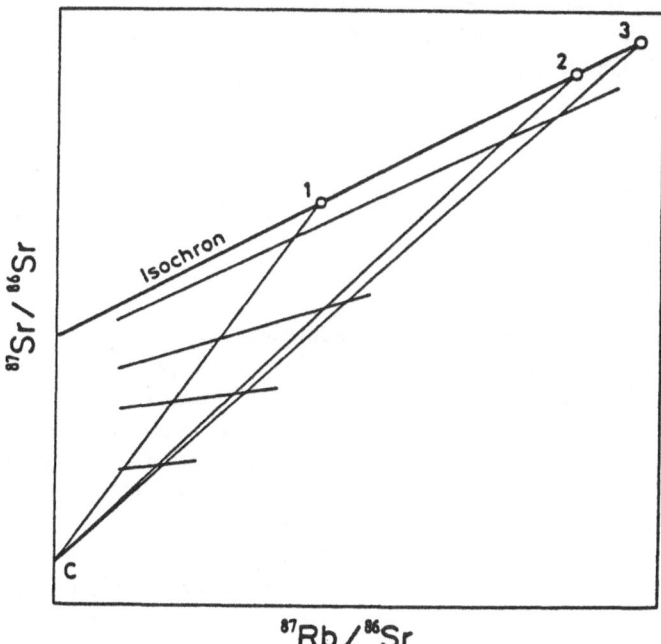

Fig. 7.13. Rb-Sr isochron diagram illustrates schematically the influence of a carbonate component (C) on the Sr isotopic composition of the authigenic clay mineral phases 1 to 3 which are in isotopic equilibrium with each other. It is assumed that authigenic and detrital phases are mixed with each other in approximately the same proportion (condition for straight line errorchron without age significance). Adding high amounts of carbonate (C) strongly lowers the slope of the corresponding errorchron (for further explanation see text).

Several possible straight lines may result from mixing according to various mixing parameters (f; Faure, 1986). The gradients of the resultant straight lines will decrease with increasing carbonate strontium influence. This effect will serve to decrease the apparent age of the samples.

As the REE are highly mobile in acidic environments and tend to precipitate under conditions of high pH (see Sect. 3.2), it can be assumed that the alkaline carbonate-rich, fluid phase only contained very low REE concentrations. Likewise, it can be assumed that the REE incorporated into, or adsorbed onto, the clay minerals were not mobilized. REE from other sources are also unlikely to have been added to the system under these conditions. This would explain how the Sr isotopic system can be opened while the Nd isotopic system may remain closed during late diagenesis. Further possibilities of this decoupling will be pursued in the next part.

Therefore, the organic matter in clastic sediments, whose degradation creates the necessary conditions for carbonate dissolution, plays an important role not only for the in situ crystallization of isotopically homogeneous clay minerals after feldspar dissolution but also for the closed system behavior of REE and Pb. These findings are important also in the context of nuclear waste disposal. Clay minerals togther with organic matter may help retain or retard actinides and fission products released from nuclear waste (Bros et al., 1993; Stille and Gauthier-Lafaye 1997; Gauthier-Lafaye et al. 1997).

7.3 Nd Isotopic Homogenization in Phosphatic, Detrital Sediments

The Sm-Nd isotopic system in detrital sediments, which have only minor amounts of organic material but detectable quantities of phosphate-rich mineral phases such as apatite, behaves quite differently than the Sm-Nd isotopic system in bituminous clays. Phosphates preferentially incorporate rare earth elements and can therefore influence the process of Nd homogenization and the behavior of the Sm-Nd isotopic system of the whole rock as has been shown by Awwiller (1994), Ohr et al. (1994) and Schaltegger et al. (1994).

Authigenic apatite forms as a result of early diagenetic processes close to the sediment-water interface (Jahnke et al., 1983; Glenn und Arthur, 1988; Van Cappellen and Berner, 1991) depending on the availability of phosphorus in the porewaters. The incorporation of rare earth elements into the initially REE depleted biogenic apatite takes place by the coupling of the Si^{4+}/P^{5+} -substitutions onto the tetrahedral positions REE^{3+}/Ca^{2+} and Na^+/Ca^{2+} onto the 2 Ca positions. It is assumed that REE can be incorporated as a result of a variety of processes. The enrichment of REE and dissolved phosphorous in the porewaters of anoxic sediments is thus brought about mainly by the decay of organic components, reduction of hydrolized Mn and Fe oxides, and by the dissolution of phosphate-rich fish bone fragments (Elderfield et al. 1981; Elderfield und Sholkovitz 1987; Sholkovitz et al. 1989). For early diagenetic apatite to form, it is particularly important that a sediment be rich in organic matter as almost all the dissolved phosphorus in porewaters is likely to come from this source. Experimental data of Van Cappellen and Berner (1991) show that authigenic phosphate has grain-sizes between 0.1 to 10 μm.

Schaltegger et al. (1994) investigated greywackes from a metamorphic profile southwest of Casablanca (Morocco). The sample material was derived from a non-metamorposed, diagenetic zone of anchi- and epi-zonal metamorphic facies. The acid leachates of the clay fractions were generally rich in Ca, Mg and Fe, which suggests dissolution of ankeritic carbonate and Fe-oxides/hydroxides. The P_2O_5 contents in the acid leachates were likewise increased and allow themselves, with one exception, to be correlated with Nd concentrations (Fig. 7.14). The close relationship between Nd and phosphate content allows us to assume that apatite is the most important carrier of the REE in these rocks. REE concentrations are enhanced in the acid leachates (1000 x chondrite concentrations) and their patterns suggest primary apatite fractionation (Fig. 7.15). Enrichment of the intermediate REE (MREE) appears to be characteristic for some types of phosphate (see also Awwiller 1994). The leachates define a straight line in the Sm-Nd isochron diagram whose gradient yields an age of 523 Ma (Fig. 7.16).

Schaltegger et al. took this age to be that of Nd isotopic homogenization between the leachable phases (apatite, Fe-oxide/hydroxide) during diagenesis.

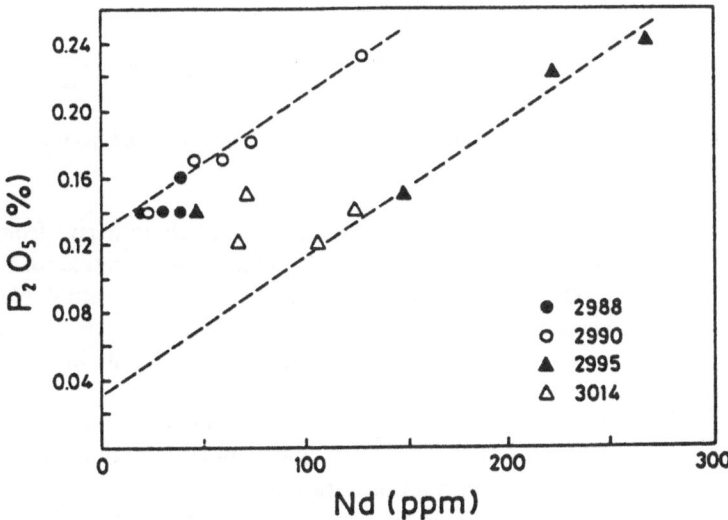

Fig. 7.14. Relationship between Nd and P_2O_5 concentrations in acid leachates of clay fractions from phosphate bearing detrital sediments. (Schaltegger et al. 1994)

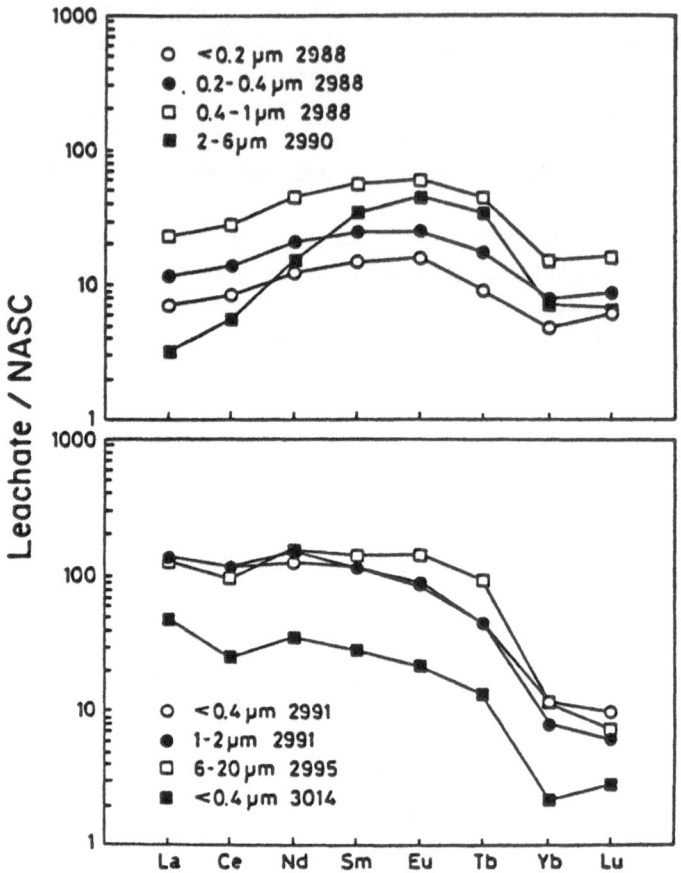

Fig. 7.15. NASC normalized acid leachates of clay fractions (<0.4 μm) from phosphate bearing sediments. (Schaltegger et al. 1994)

The variation in the Sm/Nd ratios of the leachates is likely to derive from different mixing ratios of apatite and Fe-hydroxides with high Sm/Nd ratios (0.21) and low Sm/Nd ratios (0.09), respectively, both having identical initial Nd isotopic compositions. Awwiller (1994) demonstrated that amorphous iron hydroxide grain coatings are second only in importance to phosphate minerals in controlling the leachable REE reservoir.

It can be assumed that a large proportion of Sm and Nd in the residues derives from crystallographic sites in the insoluble silicate phases. The initial $^{143}Nd/^{144}Nd$ ratios of the residue <0.4 μm fractions are almost identical to those of the acid leachates (Fig. 7.16) and allow us to suppose that the finest clay mineral fractions reflect the isotopic composition of the diagenetic fluid phase in which the apatite and the Fe-oxide/hydroxide phases crystallized. The authors make the case that diagenesis of clay-rich sediments can be dated using Sm-Nd isotopic analyses of authigenic, accessory minerals, which grew in the remaining pore spaces (cement-mineral phases). Diagenetic fluid phases were pumped through associated primary and secondary pore spaces for some distance as a result of compaction and excessive overburden, which must have led to large-scale isotopic homogenization in the migrating fluids and thus also with the crystallizing authigenic cement mineral phases. The Sm-Nd isotopic system would appear to be suitable for dating the process of diagenetic compaction and cement formation in this case.

The Sm-Nd isotopic system seems to have behaved rather differently than the Rb-Sr isotopic system in the case of Morocco, too. The only lightly metamorphically overprinted, untreated clay fractions define a straight reference line on the Sr isochron diagram, yielding an age of 309 ± 43 Ma (Fig. 7.17A). The linear relationship between $^{87}Sr/^{86}Sr$ and $1/Sr$ ratios (Fig. 7.17B) shows clearly that this reference line in Fig. 7.17A originates from the mixing of two isotopically different components. This age cannot be geologically relevant.

The Rb-Sr isotopic data for the finest <0.2 μm clay fractions, which contain to a large extent newly formed illite and little detritus, define an isochron age of 342 ± 21 Ma in Fig. 7.17C. Is this age geologically relevant? Could this be a reflection of Hercynian overprinting? Which mechanism allowed the reopening and subsequent rehomogenization of the Rb-Sr isotopic system? Schaltegger et al. discussed possible physico-chemical conditions that would have permitted a decoupling of the two isotopic systems.

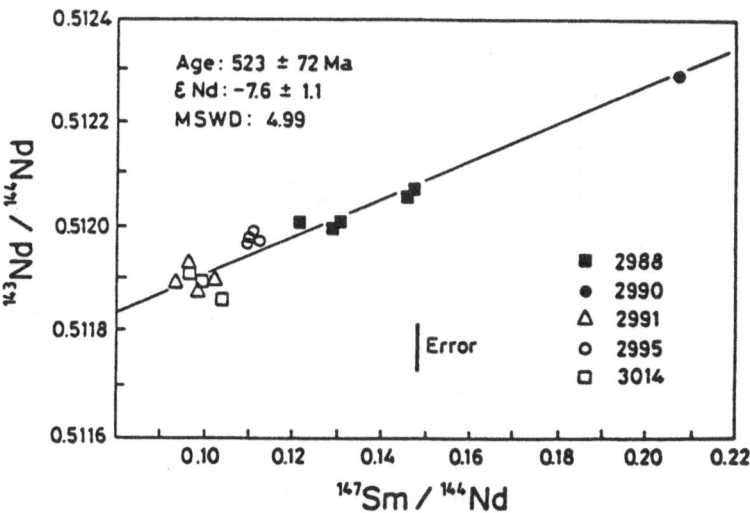

Fig. 7.16. Sm/Nd isochron diagram for acid leachates of various fractions from phosphate bearing sediments. All components go towards defining an initial age of homogenization of 523 Ma. Filled symbols: leachates; open symbols: residues. (Schaltegger et al. 1994)

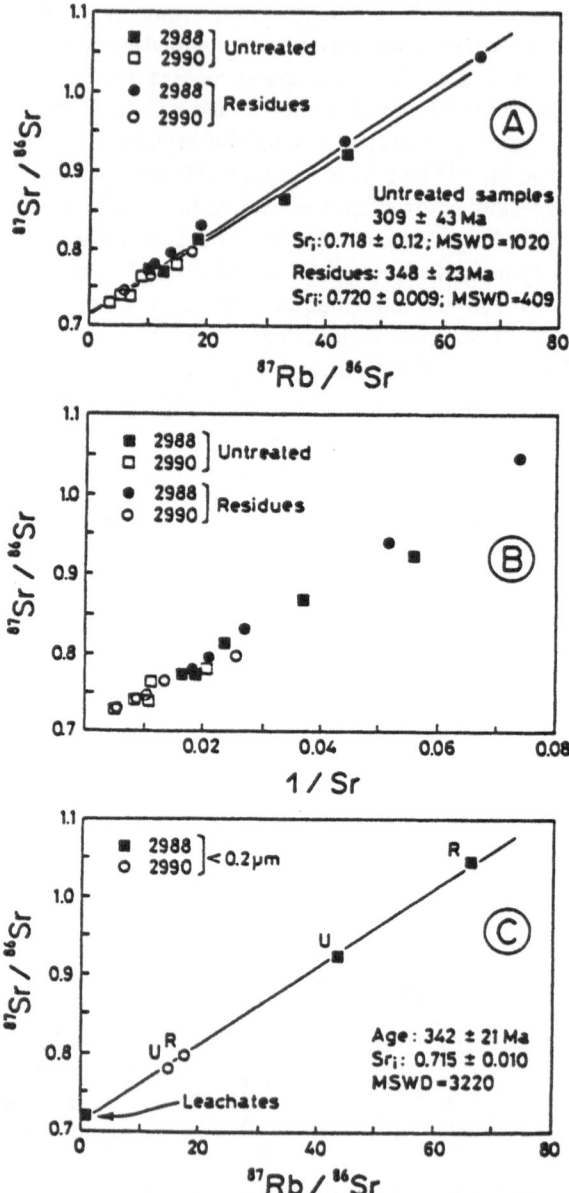

Fig. 7.17. Rb-Sr isochron and mixing diagrams for clay minerals from phosphate bearing sediments (Schaltegger et al. 1994)

They assume that the Sm-Nd isotopic system of the apatite and Fe-oxide/hydroxides remained closed during Hercynian overprinting because, 1) the water/rock ratio remained low during this overprinting compared with the ratio that existed during diagenetic compaction and dewatering, 2) the fluid phases were rich in strontium but depleted in REE, and 3) the cement minerals are strongly enriched in REE and would react less sensitively to external neodymium. The decoupling of the two isotopic systems can also be controlled by mineralogic effects such as the growth of minerals which carry various amounts of Sm-Nd and Rb-Sr and which can develop and exchange with each other independently of various thermal events. The fact that the finest clay fraction (<0.2 μm) provides an age of isotopic homogenization of 523 Ma for Nd and around 300 Ma for Sr demonstrates that the decoupling of both systems from each other is related to the different bonding of Nd and Sr with the crystal lattice of layered silicates such as chlorite and illite.

The influence of phosphate-rich phases, in particular apatite, on the development of the Sm-Nd isotopic system during diagenesis or weak metamorphism was also discussed by Ohr et al. (1994). The investigated samples allowed the authors not only to separate and analyze the clay minerals but also authigenic apatite. This apatite is characterized by a relatively high $^{147}Sm/^{144}Nd$ ratio (0.3) and an initial $^{143}Nd/^{144}Nd$ ratio, which when calculated back to the time of sedimentation or early diagenesis gives an almost identical isotopic composition to that of the clay minerals. The clay minerals and the apatite appear to have undergone early isotopic exchange with the same fluid phase at the time of their formation. thus incorporating Nd of the same isotopic composition. The Sm/Nd ratios, determined on various clay mineral fractions, show a strong grain size dependency.

In the study of Ohr et al., the Sm/Nd ratios fall with decreasing grain size. The Sm/Nd ratios of the bituminous shales (Sect. 7.2) behaved in exactly the opposite fashion. In the latter case, finer grain-size fractions had higher Sm/Nd ratios. Obviously different mechanisms must be responsible for the fractionation of Sm with respect to Nd. We assume that progressive crystallization of clay minerals and the formation of organo-metallic complexes led to REE fractionation in the case of the bituminous clay shales. Phosphate-bearing detrital sediments behave differently. The leaching experiments of Ohr et al. (1994) on clays help to demonstrate this. These authors observed that the amount of leachable Nd and Sm in the samples increased with increasing grain-size, while the Sm/Nd ratios of the leachates decreased. Provided the REE ratios and abundances are controlled by the presence of acid-soluble and REE-rich mineral phases, it can be assumed that these phases possess lower Sm/Nd ratios but higher REE concentrations in the coarser fractions than in the finer fractions. This observation was confirmed using the electron microscope: it was shown that the coarse fraction consisted largely of monazite and florencite, which are particularly enriched in the lighter REE, whereas the finer fractions contained more apatite, which tends to be enriched in the middle REE.

7.4 References

Awwiller DN (1994) Geochronology and mass transfer in Gulf Coast mudrocks (south-central Texas, U.S.A.): Rb-Sr, Sm-Nd and REE systematics. Chem Geol 116: 61-84

Bros R, Stille P, Gauthier-Lafaye F, Weber F, Clauer N (1992) Sm-Nd isotopic dating of Proterozoic clay material: An example from the Francevillian sedimentary series (Gabon). Earth Planet Sci Lett, 113: 207-218

Bros R (1993) Géochimie isotopique (Sm-Nd, Rb-Sr, K-Ar, U) des argiles du bassin protérozoique de Franceville et des réacteurs d'Oklo (Gabon). PhD Thesis, Univ. Strasbourg

Bonhomme MG, Gauthier-Lafaye F, Weber F (1982) An example of Lower Proterozoic sediments: the Francevillian in Gabon. Precamb Res, 18: 87-102

Elderfield H, Hawkesworth CJ, Greaves MJ (1981) Rare earth element geochemistry of oceanic ferromanganese nodules and associated sediments. Geochim Cosmochim Acta, 45: 513-528

Elderfield H, Sholkovitz ER (1987) Rare earth elements in the porewaters of reducing nearshore sediments. Earth Planet Sci Lett, 82: 280-288

Gauthier-Lafaye F, Bros R, Stille, P (1996) Pb-Pb isotope systematics on diagenetic clays: an example from Proterozoic black shales of the Franceville Basin (Gabon). Chem Geol (Isot Geosc),133:243-250

Gauthier-Lafaye F, Holliger P, Blanc PL (1996) Natural fission reactors in the Franceville basin, Gabon: a review of the conditions and results of a critical event in a geologic system. Geochim Cosmochim Acta, 60: 4831-4852

Glenn CR, Arthur MA (1988) Petrology and major element geochemistry of Peru margin phosphorites and associated diagenetic minerals: Authigenesis in modern organic-rich sediments. Mar Geol, 80: 231-267

Jahnke RA, Emerson SR, Roe KK, Burnett WC (1983) The present day formation of apatite in Mexican continental margin sediments. Geochim Cosmochim Acta, 47: 259-266

Manning LK, Frost CD, Branthaver JF (1991) A neodymium isotopic study of crude oils and source rocks: potential applications for petroleum exploration. Chem Geol, 91: 125-138.

Ohr M, Halliday AN, Peacor DR (1991) Sr and Nd isotopic evidence for punctuated clay diagenesis, Texas Gulf Coast. Earth Planet Sci Lett, 105: 110-126

Ohr M, Halliday AN, Peacor DR (1994) Mobility and fractionation of rare earth elements in argillaceous sediments: Implications for dating diagenesis and low-grade metamorphism. Geochim Cosmochim Acta, 58: 289-312

Peterman ZE (1966) Rb-Sr dating of Middle Precambrian metasedimentary rocks of Minnesota. Geol Soc Amer Bull, 77: 1031-1044

Schaltegger U, Stille P, Rais N, Piqué A, Clauer N (1994) Neodymium and strontium isotopic dating of diagenesis and low-grade metamorphism of argillaceous sediments. Geochim Cosmochim Acta, 58: 1471-1481

Sholkovitz ER, Piepgras DJ, Jacobsen SB (1989) The pore water chemistry of rare earth elements in Buzzards Bay sediment. Geochim Cosmochim Acta, 53: 2847-2856

Stille P, Clauer N (1986) Sm-Nd isochron-age and provenance of the argillites of the Gunflint Iron Formation in Ontario, Canada. Geochim Cosmochim Acta., 50: 1141-1146

Stille P, Gauthier-Lafaye F, Bros R (1993) The neodymium isotope system as a tool for petroleum exploration. Geochim Cosmochim Acta, 57: 4521-4525

Stille P, Gauthier-Lafaye F (1997) Pb and Nd isotopic equilibrium and closed systems in bituminous clastic sediments? 7th. Annual Goldschmidt Conference, Tucson, Arizona, June 2-6

Toulkeridis T, Goldstein SL, Kröner A, Lowe DR, Schidlowski M (1990) Late Archean Rb-Sr, Pb-Pb and Sm-Nd resetting of early Archean Barberton Greenstone Belt carbonates. In Glover, JE and Ho, SE, eds., Third International Archean Symposium (extended abstracts): Perth, Australia, Geoconferences (WA Inc), 309-310

Toulkeridis T, Goldstein SL, Clauer N, Kröner A, Lowe DR (1994) Sm-Nd dating of Fig Tree clay minerals of the Barberton greenstone belt, South Africa. Geology, 22: 199-202

Van Cappellen P, Berner RL (1991) Fluorapatite growth from modified seawater solutions. Geochim Cosmochim Acta, 55: 1219-1234

Subject Index

—A—

Aare 71
acid rain 11; 30; 48
activity 3; 64; 77-83
adsorption 21; 42; 53
aerosols 53-54; 148
air filter 59
allanite 56
anthropogenic 47-49; 55-56; 59-63; 72;
   89; 135
apatite 25; 56; 102; 122-123; 175; 188;
   192; 204-209
apparent age 16; 162-163; 166-167; 169-
   172; 180-181; 204
Ar 16; 188
Ar loss 16
aragonite 108; 120
Argon 2; 6; 162
Atlantic 55; 96; 109; 112-124; 128-133;
   135; 137-141; 155; 166
Atlantic Ocean 120
atmosphere 10-11; 47-49; 53-57; 63; 76;
   83; 186
authigenic 11; 18; 89; 94; 96; 102; 114;
   136; 139-140; 143-144; 156-157;
   164-168; 176-182; 185; 188; 192;
   195; 197; 200-209

—B—

Banded Iron Formations 126
barite 94
becquerels 64
beer 59
biogenic calcites 120
biogenic phosphate 122
biosphere 11; 47; 52; 76
biostratigraphy 94; 145
biotite 11; 14-20; 25-30; 168; 189-190;
   201
bitumen 190; 196-201
black shales 146; 188; 190; 202
blood 59
bottom water 133; 135
brachiopods 94; 122; 124; 127
Broken Hill lead 50
Broken-Hill 59
Bushveld complex 62

—C—

Cambrian explosion 105-106
carbon 1; 2; 43; 56-59; 76; 102; 160;
   188; 190; 196; 198; 200
carbon budget 76
carbonate 33; 36; 40; 76; 90-94; 97-103;
   106-111; 119-123; 128; 144-145;
   159; 176; 203
carbonate diagenesis 202
carbonate dissolution 92

Lecture Notes in Earth Sciences